.75

CONTRACTOR'S GUIDE TO

QuickBooks Pro® 2008

Karen Mitchell

Craig Savage

Jim Erwin

Craftsman Book Company
6058 Corte del Cedro / P.O. Box 6500 / Carlsbad, CA 92018

Acknowledgments

The authors wish to express their thanks to *Nancy Carriere,* without whose assistance this book would never have been written.

The Library of Congress has cataloged the prior edition as follows:

Mitchell, Karen, 1962-
  Contractor's guide to QuickBooks Pro 2007 / Karen Mitchell, Craig Savage, Jim Erwin.
    p. cm.
  Includes index.
  ISBN-13: 978-1-57218-186-1 (alk. paper)
  1. Construction industry--Accounting--Computer programs. 2. Construction
industry--Finance--Computer programs. 3. Building--Estimates--Computer
programs. 4. QuickBooks. I. Savage, Craig, 1947- II. Erwin, Jim, 1943-
III. Title. IV. Title: QuickBooks Pro 2008.

  HF5686.B7M4934 2008
  657'.8690285'53--dc22

                          2006102995

© 2008 Craftsman Book Company
ISBN-13: 978-1-57218-200-4

Text update edited by Maria Mercedes Strong
National Estimator instructions (Appendix B) updated by Lori Boon

Cover design by John Wincek

Contents

Introduction

Why You Need This Book

Contractor's Guide to QuickBooks Pro is a simple, hands-on guide for contractors, remodelers, subcontractors, and real estate developers who plan to use, or are now using, *QuickBooks Pro* accounting software. Think of this manual, like QuickBooks Pro, as a powerful new tool. Used correctly, it will get results you never thought possible — in ways you never imagined.

This book was written *for* contractors *by* contractors. That's why it's low on "accounting speak" and high on practical examples. We're not going to talk about debits and credits. QuickBooks Pro doesn't use those terms, so we won't either. Instead, we'll use words we all understand, like checks, estimates, bills, timecards, purchase orders, and deposit slips.

We'll help you set up a good, effective, highly-professional bookkeeping system as quickly and painlessly as possible. You'll know, at the click of a mouse button, who owes you money and who you owe. You'll know if there's enough cash on hand to pay bills. When you need a current profit and loss statement, you'll have one in a minute or two — either for the month, the year, or for just one job. When a lender or a bonding company needs a balance sheet, you'll get one in minutes.

Contractors have special payroll requirements. This book will show you how to get the most out of QuickBooks Pro's payroll system. No matter where you do business in the 50 states, QuickBooks Pro has a current tax table exactly right for your company. If QuickBooks Pro isn't doing your payroll now, we predict it will be before too long.

Since estimating is important to most construction contractors, we're going to cover estimating from three perspectives. In Chapter 12, you'll see how to enter the summary data from estimates you've already written into QuickBooks Pro. From here, you can do progress billing and job cost tracking. In Appendix A you'll learn how to use the estimating system built into QuickBooks Pro to create estimates. And for those who want still more estimating power, Appendix B has complete instructions for using Craftsman's estimating software, *National Estimator*, and *Job Cost Wizard*, a program that lets you turn estimates into invoices that will export to QuickBooks Pro. Both are included on the CD in the back of this book.

What You Can Expect from This Book

According to a recent national survey, more construction contractors use QuickBooks Pro and QuickBooks than all other accounting programs combined. When set up properly, QuickBooks Pro can handle the accounting for most small- to medium-sized (to $10 million a year) construction companies. But despite the sales hype, QuickBooks Pro isn't easy to set up and learn. Dozens of options and preference settings may lead you down the wrong road, resulting in hours of frustration. We've spent hundreds of hours testing the options and preferences in QuickBooks Pro to help you get it right the first time. Follow our examples and you'll have an effective accounting system that provides all the information any successful business needs — and in the shortest time possible.

QuickBooks Pro doesn't replace accountants. But it does help organize and standardize your bookkeeping system. Every report your accountant needs is readily available. This makes it easier for your accountant to prepare tax returns and annual reports. The time saved should translate into lower costs for your company.

In this book, we'll explain how to:

- set preference options correctly for your company
- set up a Chart of Accounts that matches the work you do
- set up, edit, and use classes
- set up customers and jobs
- set up vendors and subcontractors
- set up employee payroll
- get your current account balances into QuickBooks Pro
- track transactions through QuickBooks Pro
- create and use estimates
- set up a simple and effective job cost system
- create and send invoices
- enter vendor bills
- write checks
- process payroll

- get payroll tax and workers' comp expenses into job cost reports

- get an owner's time into job cost reports (for a sole proprietorship)

- run workers' comp reports

- create and interpret job cost reports to keep track of your business

- prepare financial statements

- set up end-of-the-month and end-of-the-year procedures

In short, we'll show you how to get everything a construction company needs out of QuickBooks Pro. If you understand and apply the methods in this book, you should see real improvement in the effectiveness of your accounting system. And, you'll gain the personal and financial rewards that come from working not just harder, but smarter.

QuickBooks Simple Start, QuickBooks Pro or QuickBooks Premier Contractor Edition?

You'll notice that the title of this manual refers to the "Pro" version of QuickBooks. We recommend that you use at least QuickBooks Pro because QuickBooks Simple Start can't create estimates and can't track time spent on specific jobs. If time tracking and estimating are important to your construction company, invest the extra money to get QuickBooks Pro. If you have QuickBooks Premier Contractor Edition, this book will work for you as well. All the features in this book are based on QuickBooks Pro, which is included in the Contractor edition. When we refer to QuickBooks Pro, we are also referring to QuickBooks Premier Contractor Edition.

You'll find another major advantage to QuickBooks Pro once we get into estimating. QuickBooks Pro can do progress billing — creating an invoice for each part of a job as you complete that part. That's important if you handle larger jobs that take weeks or even months. For example, you can send out a bill that covers 100 percent of the foundation work, 40 percent of the framing, and 20 percent of the plumbing on a job. That's an important advantage. No calculation is required. QuickBooks Pro does the math and keeps all the records: what you told the customer it would cost, what you've billed so far, and what is left to be billed. These progress billing statements may also become important business records to reduce arguments over what's still owed.

We believe QuickBooks Pro has the best combination of power and simplicity for small-volume builders and general contractors. It makes accounting simple for non-accountants because it works the way you do. You've been writing checks for years. With QuickBooks Pro, you fill in blanks on a check the same as always. But you do it on a computer screen.

And you add notes to check stubs about accounts and jobs to be charged — just the way you've always added notes to the stubs of paper checks. Timecards, purchase orders, and invoices work the same way.

Can QuickBooks Pro do the job for you? The thousands of construction companies now using QuickBooks Pro are good evidence that it can. If you're serious about making your construction business grow and prosper, you'll want the accounting and reporting power built into QuickBooks Pro.

There's an old saying among builders about construction accounting: "A builder who knows where he stands won't stand there very long." You need to know where you stand so you can make informed decisions quickly. If you agree, QuickBooks Pro may be perfect for your company.

For more than ten years, the three authors of this book have used construction accounting packages, some costing thousands of dollars. None are as slick, professional, and as easy to use as QuickBooks Pro. In our opinion, QuickBooks Pro offers the builder, remodeler, general contractor, and specialty subcontractor the best off-the-shelf accounting program on the market. It's affordable, reliable, and probably has all the features you'll need.

Even though QuickBooks Pro never mentions debits and credits, it handles accounting and reporting functions the same way an accountant would. It follows what professional bookkeepers and accountants refer to as Generally Accepted Accounting Principles (GAAP). You may never notice, but QuickBooks Pro uses conventional double-entry accounting. That means each time you enter a transaction, the numbers go two directions — one way as a debit and another way as a credit. Suppose you enter a bill from a supplier. Behind the scenes, QuickBooks Pro records the transaction two ways, first as an account payable and then as a charge to an expense category.

Strictly speaking, the way it comes out of the box, QuickBooks Pro isn't a true construction accounting program. But it's adaptable enough to fit the needs of most construction companies like a glove. For example, you can customize reports to get great job cost reports, just like a so-called "construction accounting package." That's why we feel so strongly that QuickBooks Pro is right for most construction companies.

Why Should You Believe Us?

All three authors have been in construction, using computers, and using QuickBooks and QuickBooks Pro for many years. We've helped hundreds of contractors set up and use QuickBooks and QuickBooks Pro. We're confident that what we've done for others we can do for you, too.

Karen Mitchell was a general building contractor and is currently co-owner of Online Accounting (www.onlineaccounting.com) which uses the Internet to help train contractors on QuickBooks Pro and Master Builder. Karen conducts seminars nationwide for contractors who use QuickBooks Pro. She is a frequent speaker at many construction trade shows such as: A/E/C SYSTEMS, JLC Live!, and NAHB's PCBC (Pacific Coast Builders Conference). Karen has written several books including: *Construction Forms & Contracts, Quicken for Contractors, Contractor's Guide to QuickBooks Pro, Architect's Guide to QuickBooks Pro, Interior Designer's Guide to QuickBooks Pro*, and *Real Estate Investor's Guide to QuickBooks Pro*.

Craig Savage has been a general building contractor, remodeler and custom homebuilder for over 25 years. He was an editor at *The Journal of Light Construction* magazine for many years, director of the *JLC LIVE! Training Shows*, Vice President of Marketing & Sales at www.BobVila.com, and most recently, VP of Marketing at Building Media, Inc.

In his free time Craig is a construction management computer consultant. He started *Construction Business Computing* and *Macintosh Construction Forum* newsletters, and his articles have appeared in *Architectural & Engineering Systems, Architectural Record, Fine Homebuilding Magazine, Computer Applications Newsletter, Remodeling News, NAHB Commercial Builder, NAHB Single Family Forum, Remodeler Magazine, Mac Week, Document Imaging, Imaging World*, and *A/E/C Computer Solutions*.

Craig is a regular speaker at the A/E/C Systems, NAHB, NARI, CSI, and PCBC annual conventions. He also instructs at seminars sponsored by the University of Wisconsin College of Engineering, and the University of California Santa Barbara Extension.

Other books he has co-written for Craftsman Book Company are *Construction Forms & Contracts* and *Quicken for Contractors*. With Taunton Press he wrote *Trim Carpentry Techniques*.

Jim Erwin is a partner in several second-generation family-owned construction companies in upstate New York that are involved in land development as well as residential and light commercial construction. He's an active member of the National Association of Home Builders and has written articles on using computers in construction for a variety of construction magazines. He's also the creator of GC/Works (published by sYnapse Software, Inc.), a full-featured software solution for the construction industry that uses Quicken or QuickBooks Pro as its basis.

What Comes Next?

But don't just take our word for it. Here's a brief summary of some of the features we'll show you to make QuickBooks Pro specific for your construction business needs.

Payables/Receivables

One of QuickBooks Pro's greatest strengths is in tracking money owed to you (Accounts Receivable) and money you owe others (Accounts Payable). For example, when you buy materials from a vendor and receive an invoice or delivery slip, you enter the information in a screen form called a Bill. You've seen lots of these, even if you've never used a computer and don't know anything about accounting.

QuickBooks Pro tracks payables and receivables so you can see at a glance what's owed and what's due. You can even see what's due on several different reports: Accounts Receivable Aging, Open Invoices, and Customer Balance Summary. Having all this information available almost instantly should help you sleep better at night. And it helps you make better decisions. For example, it's nice to know how long it's been since you received the last check when a customer asks for "just a little more work" on a project.

The Estimating Programs and Cost Data

QuickBooks Pro doesn't come with cost estimating data. And the estimating function built into QuickBooks Pro is limited, as you'll discover in Chapter 12. That's why we include the National Estimator and Job Cost Wizard programs. You can create an estimate with National Estimator and then import it into QuickBooks Pro. Everything you need for estimating is on the CD inside the back cover of this book and gets installed when you select the "Complete" installation.

Purchase Orders

There are three good uses for the purchase orders in QuickBooks Pro. The first is to track orders placed for special order items, such as windows, doors, and skylights. Another is to track what you've committed to pay subcontractors. For example, suppose a subcontractor bids a project at a certain price. You can use a purchase order to track the bid price to compare it with the bill he submits. The third use is to establish a schedule for multiple payment draws for a subcontractor. If you agreed to pay a framing contractor, say, 30 percent on completion of the floor framing, 30 percent on completion of the walls, and 40 percent after the roof framing, you can use purchase orders to keep track of those payment schedules and payments you make against them.

Time Recording and Payroll

Payroll and time billing are flexible and sophisticated functions in QuickBooks Pro. The timecard window looks like a paper timecard, so data entry will be a familiar task. You'll find QuickBooks Pro's payroll function to be very accommodating.

QuickBooks Pro payroll handles tax withholding, additions, deductions, and company contributions. Deductions such as health insurance, payments to a retirement plan, or union dues are simple to set up and report. So are company contributions such as health, life, and workers'

Intro-1

A National Estimator estimate imported into QuickBooks Pro using Job Cost Wizard.

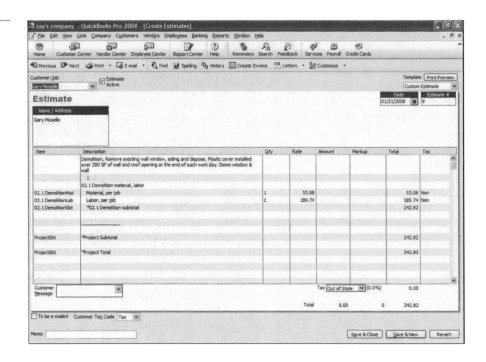

comp insurance paid by the company. You can record irregular payroll events such as bonuses, reimbursed travel expenses, and advances against salary. QuickBooks Pro even keeps track of sick and vacation time due employees.

QuickBooks Pro payroll reports provide all the information you need to file state and federal employer tax forms in any of the 50 states. That's part of the QuickBooks Pro payroll system. Intuit, the developers of QuickBooks Pro, have a small additional charge for the current tax table. Payroll tax tables are updated automatically as tax rates change.

If you want, you can use the QuickBooks Pro Online Payroll Service to pay all of your state and federal taxes and file all the necessary forms (including W2s) for a modest cost. You can have QuickBooks Pro print payroll checks, or use the QuickBooks Pro Online Direct Deposit Service to deposit employee paychecks automatically at nearly any bank.

What's on the CD?

We hope you bought this book because there's a CD in the back. We've loaded the CD with everything you need to get the most out of QuickBooks Pro:

▌ A QuickBooks Pro data file with a Chart of Accounts, items list, class list, and memorized reports for you to adapt to your business. Just plug in your own company data — vendors, subs, customers, etc. — and you're up and running.

▮ *National Estimator*, an easy-to-use estimating program with over 200 pages of construction cost estimating data for general contractors. The subtotals you create in National Estimator become cost categories in QuickBooks Pro.

▮ *Job Cost Wizard*, software that converts your National Estimator estimates into QuickBooks Pro format so you can create invoices and track costs on every job. (Requires Windows 95 or later.)

▮ *ShowMe*, a 60-minute interactive video that shows how to use National Estimator, Job Cost Wizard, and QuickBooks Pro to estimate and track job costs.

▮ Sample business forms in the Forms folder. You'll find a grid to track change orders and a timecard to collect payroll data from tradesmen in the field. Other forms will help you organize information gathering in your office. Estimators may like using the estimate summary form to collect estimating data for transfer to QuickBooks Pro. To use these forms, you'll need a word processing or spreadsheet program such as Microsoft Excel, Microsoft Word, Microsoft Works, WordPerfect, or Lotus 1-2-3.

However, the CD in the back of this book doesn't include QuickBooks Pro. You won't get much out of this book without a working copy of QuickBooks Pro. So the first step will be buying and installing QuickBooks Pro if you don't have it already.

Fortunately, you can get QuickBooks Pro at most large software outlets. For the lowest price, try searching for a dealer on the Web. Some dealers offer expedited delivery at little or no cost. The CNET site lists dozens of merchants who offer QuickBooks Pro. To see the current price and terms offered by each of these merchants, go to *http://shopper.cnet.com/*.

How Do I Use This CD?

To use the CD in the back of this book, you'll need a computer running Windows XP or Vista, a CD-ROM drive, and 80 - 100MB available on your hard drive.

If you're using a version of QuickBooks Pro older than the 2008 version, many of the illustrations in this book may not look exactly like what's on your screen. That's because this manual is based on QuickBooks Pro version 2008. If you're using version 2000 - 2007, the changes will be mostly cosmetic. Version 99 and older versions are different.

Intro-2

Select the correct disk letter and type setup.

Installing the CD

To install everything on the CD, put the Contractor's Guide to QuickBooks Pro CD in your CD drive (such as D:). If installation doesn't start automatically after a few seconds:

▌ Click **Start** button

▌ Click **Run**

▌ Enter **d:\setup.exe**

Then follow the instructions on the screen. We recommend you select the "Complete" installation when you're asked which type of installation you want to use.

When installation is complete, you'll see a new program group when you click Start and then click Programs. The new program group name is Construction Estimating.

Get Help by Phone

Everything you need to know about National Estimator and Job Cost Wizard is available on the Help files that come with each program. Just click Help on the menu bar or click the question mark at the right end of the toolbar, or press the F1 key.

If you have trouble installing or using National Estimator or Job Cost Wizard, call Craftsman Book Company (Monday through Friday from 8 a.m. to 5 p.m. Pacific time) at 760-438-7828.

If you need help with QuickBooks Pro, call Intuit technical support (Monday through Friday from 6 a.m. to 4 p.m. Pacific time) at 888-320-7276 or call Online Accounting (Monday through Friday from 8 a.m. to 5 p.m. Mountain time) at 888-254-9252.

Removing the Installed Files

To remove any of the programs installed from the Contractor's Guide to QuickBooks Pro CD from your hard drive:

▌ Choose **Start**, **Control Panel**, then click on **Add or Remove Programs**.

▌ Click the name of the program you want to remove, **Contractor's Guide to QuickBooks Pro 2008**.

▌ Click **Change/Remove**.

▌ Click **Yes**.

Conclusion

Most contractors would agree that accounting is what they like least about running a construction business. When we started out, we felt the same way. We had a well-founded fear of accounting and an irrational loathing of computers. But using a computer for the first time isn't much different from using a Skilsaw for the first time. Treat it with respect. You'll gain confidence with every use.

We wrote this book because so many of our friends and colleagues asked us for a simple guide to setting up a construction accounting system. We've worked hard to keep it simple and still provide all the information you need. We feel the mission has been accomplished and hope you agree.

Now it's time to take the plunge. In Chapter 1, we'll wade right in by giving you some choices on how best to start using QuickBooks Pro for your company accounting.

Setting Up Your QuickBooks Pro Company

Getting Started

Before moving forward, you should have both QuickBooks Pro 2008 and the CD in the back of this book installed on your computer.

We're not going to include any instructions for installing QuickBooks Pro. Intuit has done a pretty good job of that. But the disk in the back of this book is our product. So here's what you need to know about installing from the *Contractor's Guide to QuickBooks Pro* disk.

1. Insert the disk in your CD drive.

2. After a short introduction, the installation screen will appear.

3. Click **Install Software** and follow the instructions on the screen. We recommend accepting the default values.

If installation doesn't start automatically, click the **Start** button. Click **Run**. In the Open field, enter your CD-ROM drive (typically D: or E:) followed by setup.exe. For example, if your CD drive is D, enter D:\setup.exe. (See Figure Intro-2) Click **OK**.

Once installation from the *Contractor's Guide to QuickBooks Pro* disk is complete, you'll probably want to start QuickBooks Pro. Click the **QuickBooks Pro** icon on your desktop.

QuickBooks Pro Company Files

QuickBooks Pro keeps all of your company records on disk in a single file. QuickBooks Pro refers to this file as your "company" file and that's what we'll call it in this manual. For many users, the first task is setting up a company file using QuickBooks Pro's "EasyStep Interview." We've designed a better way, as you'll soon see. And we recommend that you try it our way.

You can set up as many company data files as you want. The only requirement is that each one must have a different file name. We recommend that you use our *sample.qbw* file for practice while you experiment with QuickBooks Pro. When you've gained enough confidence to take off the training wheels, you can start your own "real" company with real records in the Company file.

Storing all company records in a single file simplifies moving your QuickBooks Pro company from one computer to another. A utility program built into QuickBooks Pro makes it easy to create a backup of any company file. We'll have more to say on making backups and moving your company file later in this book.

How to Find Your Company Data File

We have created the company data file and some other sample files to get you started off on the right foot. Regardless of where you have installed Intuit's QuickBooks Pro or The National Estimator, the *Contractor's Guide to QuickBooks Pro* setup program will install the example files to C:\Program Files\Craftsman.

To search for your QuickBooks folder:

▌ Click **Start**, **Search**, then **All Files and Folders**.

▌ Enter QBW32.EXE, and click **Search**. Under "In Folder" you'll see the location of QBW32.EXE. Remember that folder. You may need to find it occasionally.

To find all the QuickBooks Pro company data files on your hard drive:

▌ Click **Start**, **Search**, then **All Files and Folders**.

▌ Enter *.QBW, and click **Search**. Under "In Folder" you'll see the name and location of all company files on your hard drive.

Four Choices

Where you go from here depends on your preference and what accounting program you're using now. Here are the possibilities:

▌ You're new to QuickBooks Pro: We suggest you use our preformatted *company.qbw* file. Start at Section 1 below — Begin With Our Sample and Company Data Files.

▌ You're using a prior version of QuickBooks Pro: Begin at Section 2 — Upgrading an Old QuickBooks Pro Company to Version 2008.

▌ You're using Quicken (another product from Intuit): Begin at Section 3 — Convert a Quicken Company Data File to QuickBooks Pro.

▌ You've been using QuickBooks Pro 2008 and want to set up your company data file to match our suggestions: Begin at Section 4 — Converting an Existing QuickBooks Pro 2008 Company Data File to Our Setup.

Section 1 Begin With Our Sample and Company Data Files

We've created a sample company data file that has all the elements of a real QuickBooks Pro construction company. We call the file *sample.qbw*. It includes:

▌ a Chart of Accounts for a sole proprietorship

▌ a list of items

▌ sample customers and jobs

▌ sample vendors

▌ a list of classes

▌ payroll items set up to track workers' compensation costs and help with the workers' comp report

▌ memorized transactions

▌ memorized reports

You can use any sample file for practice while you're learning how QuickBooks Pro works. Of course, we think you'll find our *sample.qbw* data file the best one to use to learn QuickBooks Pro. We've designed it specifically for the construction industry.

However, don't use sample data files for your actual company data. When you're ready to begin entering actual records, use the company file we have on the CD called *company.qbw*. This file includes:

▌ a Chart of Accounts

▌ items

▌ payroll items

▌ classes

▌ memorized reports

We recommend modifying the company data file on the CD called *company.qbw* rather than creating your own company data file from scratch.

Besides saving time, *company.qbw* is structured to prevent errors and make it easier for you to enter your own company information. To begin:

▌ Start QuickBooks Pro.

▌ From the **File** menu, choose **Open or Restore Company** or click **Open or restore an existing company** in the No Company Open window.

▌ Select **Open a company file**, as seen in Figure 1-1. Click **Next**.

▌ Select *company.qbw* (the .qbw extension may not display, depending on your computer configuration), then click **Open**.

If you don't see *sample.qbw* and *company.qbw* in the list of QuickBooks Pro company data files, you'll need to copy them from the CD to your QuickBooks Pro folder:

▌ Put the CD from inside the back cover of this book in your CD drive. The CD will start automatically. Click **Skip**. From options, select **Exit**.

▌ Right click **Start** and select **Explore**.

▌ Scroll down the Folders side of the Windows Explorer window until you see the CD drive, probably D or E.

▌ Click the **+** to the left of the CD symbol to show all the folders on the CD.

▌ Click directly on the folder named **Examples**.

▌ Under Name on the right side of the Windows Explorer window, you'll see names of files in the Examples folder. Right click on the file you want to copy (*sample.qbw* or *company.qbw*) and select **Copy**.

▌ Now you can use Windows Explorer to find the QuickBooks Pro folder. When you've located it, right click on the folder name and select **Paste**.

▌ Right click on the name of the file you just pasted and select **Properties**.

▌ At the bottom of the Properties window, make sure the Read-only attribute is NOT checked. If it is, uncheck the box. Click **OK**.

▌ Start QuickBooks Pro.

▌ From the **File** menu, choose **Open or Restore Company** or click **Open or restore an existing company** in the No Company Open window.

▌ Select **Open a company file** (Figure 1-1).

▌ Select the *company.qbw* file.

▌ Click **Open**.

Figure 1-1

The Open or Restore Company window.

First, let's add your company information to the new file:

❚ From the **Company** menu, choose **Company Information**.

Then fill in your company information:

❚ **Company Name** — Enter your business name.

❚ **Address** — Enter the address you want QuickBooks Pro to print on invoices and purchase orders.

❚ **Legal Name** — If you registered your business with a name different from the name in the Company box, enter it here. For example, if you're incorporated as A. C. Company but do business as Any Construction Company, enter the incorporated company name here.

❚ **Legal Address** — Enter the address you want QuickBooks Pro to print on legal forms.

❚ **First month in your fiscal year** and **First month in your tax year** — Enter your company's first accounting month. Usually this is January.

❚ **Income Tax Form Used** — This depends on the type of ownership of your business. Use the drop-down list here to choose the tax form for your business's type of ownership. For example, if you own the business with someone else and it hasn't been incorporated, select Form 1065 (Partnership). If you own the business yourself and it hasn't been incorporated, select Form 1040 (Sole Proprietor). If the business has been incorporated as a regular C corporation, select Form 1120 (Corporation) or Form 1120S for an S corporation.

Figure 1-2
The No Company Open
window.

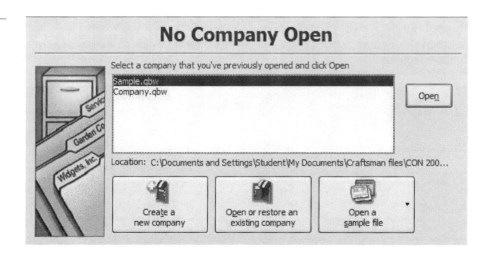

■ **Federal Employer Identification No.** - Enter your company's EIN if you'll be processing payroll.

■ Click **OK**.

■ From the **File** menu, choose **Close Company**.

It's important to note here that you've changed the company name that appears on the title bar of QuickBooks Pro. But you haven't changed the name of your QuickBooks Pro company data file. It's still *company.qbw*.

To change the *company.qbw* data file name:

■ Open the QuickBooks Pro folder.

■ Right click on *company.qbw*.

■ From the pop-up menu, choose **Rename**.

■ Enter the file name you prefer over the *company.qbw* data file name. That's the new file name of your company.

To use our *sample.qbw* data file:

■ From the **File** menu, choose **Open or Restore Company** or click **Open or restore an existing company** in the No Company Open window. See Figure 1-2.

■ Select **Open a company file**, as shown in Figure 1-1.

■ Click **Next**.

■ Select *sample.qbw* and click **Open**.

Now you're ready for Chapter 2, where we set up company preferences. You can skip the remainder of this chapter.

Figure 1-3

Check box to initiate QuickBooks update.

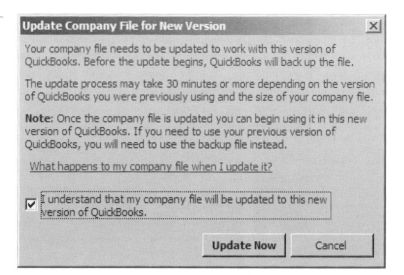

Section 2

Upgrading an Old QuickBooks Pro Company to Version 2008

To update your QuickBooks Pro company file:

▌ Install and start QuickBooks Pro.

▌ In the No Company Open window, click on **Open or restore an existing company.**

▌ Select **Open a company file**. Refer to Figure 1-1.

▌ Click **Next**.

▌ Select your existing QuickBooks file and click **Open**.

▌ At the Update Company File for New Version window, put a check in the box to confirm you want to update the file. Click **Update Now** (Figure 1-3).

▌ You'll get a message prompting you to back up your data. Click **OK**. If QuickBooks Pro doesn't display that message, it isn't set to open your old company file.

▌ At the Save Copy or Backup window, select **Local backup** (Figure 1-4).

▌ Click **Next**.

▌ In the Backup Options window, Figure 1-5, click **Browse**.

▌ Select a location to store the backup file. Typically, this will be a CD-RW or USB flash drive. Make your selection and click **OK**.

▌ Click **OK** in the Backup Options window.

▌ In the Save Backup window, click **Save**.

Figure 1-4
The Save Backup Copy window.

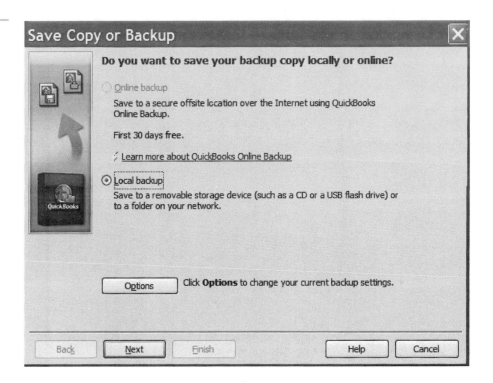

Figure 1-5
Use the Backup Options to specify the device for backup.

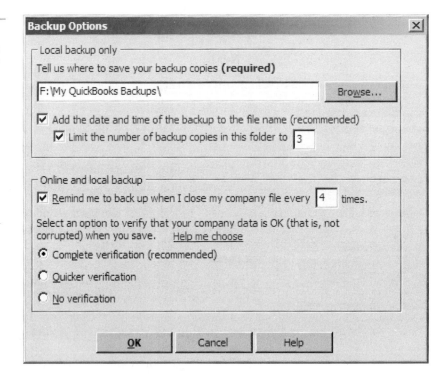

QuickBooks Pro will display a message that it's updating your data. The updating program scans your data three times. The process may seem to get stuck at a certain percentage. If there's still activity on your hard drive, don't worry. Let the update continue.

Before the file opens, confirm that you want to update the file by clicking **Yes**.

When your company has been updated, you can work through the rest of the book to set up your company data. Chapters 2 through 9 will walk you through setting up QuickBooks Pro using our methods.

If your company is different from what we recommend, you may want to follow what we recommend anyway. For example, in Chapter 3 you may find that your Chart of Accounts is different from our sample Chart of Accounts. Compare your Chart of Accounts to ours to find the differences. Add new accounts where needed and consider making inactive any accounts that aren't on our recommended list. You can inactivate accounts, items, payroll items, classes, customers and jobs, vendors, and employees. Anything marked inactive isn't deleted. It just doesn't show up on the list.

To make an account, item, payroll item, class, customer and job, vendor, or employee inactive:

▌ Open the appropriate list (i.e. Lists menu, or Customer, Vendor, or Employee Center).

▌ Select the name of the item in the appropriate list.

▌ Right click the item and select **Make Inactive**.

To see everything on the list, including anything inactive:

▌ Click **Include Inactive** at the bottom of the List window.

▌ For customers, vendors and employees, click the appropriate center. In the drop-down **View** field, select **All.**

Now you can skip the rest of this chapter and begin by setting up your QuickBooks Pro preferences in Chapter 2.

Section 3

Convert a Quicken Company Data File to QuickBooks Pro

Quicken is a check register system. QuickBooks Pro is true double entry accounting. That's a big jump. So don't expect conversion to go flawlessly. When the automatic conversion process is complete, your company data file will still need some work. That's because Quicken and QuickBooks Pro have different "buckets" for holding information. Quicken and QuickBooks Pro are both Intuit products, but they're very different, especially in the setup procedure.

Before you begin conversion from Quicken to QuickBooks Pro, make a copy of the Quicken file for safekeeping. This is very important because you'll be making changes to the data before converting to QuickBooks Pro. Convert the copy, not your real Quicken file.

▌ Start Quicken and open your Quicken data file copy. Delete any accounts that you don't want to convert (any personal or investment accounts, for example).

▌ Put any memorized transactions you want to convert into a memorized transaction group. To find out more about how to do this, see your Quicken User's Manual.

If you've entered any customer invoices as post-dated deposits:

▌ Create and print a report of those post-dated deposits. You'll use this list later to re-enter as new invoices in your new QuickBooks Pro company data file.

▌ Delete the post-dated deposits from the Quicken checking register.

Now you're ready to close the Quicken data file and work in QuickBooks Pro:

▌ Close the modified Quicken file.

▌ Start QuickBooks Pro.

▌ Choose **Close Company** from the **File** menu to close any open company data file.

▌ From the **File** menu, choose **Utilities**, then **Convert**, and then **From Quicken**.

▌ In the Important Documentation window, select **Convert**.

▌ In the Convert a Quicken File window, click the modified Quicken file, and then click **Open**. See our example in Figure 1-6.

▌ Enter a name for your new QuickBooks Pro file. In our example shown in Figure 1-7, we call the converted file mycompany.

▌ Be sure you save the file in your QuickBooks Pro directory, not the Quicken directory.

▌ Click **Save**.

At this point, you should carefully check the new QuickBooks Pro company data file. We've included information on the topics you should go through here. Make changes and additions as necessary.

Figure 1-6

Use this window to specify which Quicken file you want to convert to QuickBooks Pro.

Figure 1-7

Use this window to specify the name of a new company data file.

Company Information

To change company information:

▌ From the **Company** menu, choose **Company Information**.

Then fill in your company information:

▌ **Company Name** — Enter your business name.

▌ **Address** — Enter the address you want QuickBooks Pro to print on invoices and purchase orders.

▌ **Legal Name** — If you registered your business with a name different from the name in the Company box, enter it here. For example, if you're incorporated as A. C. Company but do business as Any Construction Company, enter the incorporated company name here.

▌ **Legal Address** — Enter the address you want QuickBooks Pro to print on legal forms.

▌ **First month in your fiscal year** and **First month in your tax year** — Enter your company's first accounting month. (Usually this is January.)

▌ **Income Tax Form Used** — This depends on the type of ownership of your business. Use the drop-down list here to choose the tax form for your business' type of ownership. For example, if you own the business with someone else and it hasn't been incorporated, select Form 1065 (Partnership). If you own the business yourself and it hasn't been incorporated, select Form 1040 (Sole Proprietor). If the business has been incorporated as a C corporation, select Form 1120 (Corporation) or Form 1120S for an S corporation.

▌ **Federal Employer Identification No.** — Enter your company's EIN if you'll be processing payroll.

▌ Click **OK**.

Customers and Jobs

When you upgrade from Quicken to QuickBooks Pro, all of the names and addresses from Quicken usually get dumped into Customers:Jobs. So you may find not only customers, but also vendors and other names in this list. If you notice this problem, first print the customer list in QuickBooks Pro:

▌ Click **Customer Center**.

▌ Click **Print** and select **Customer & Job List**.

To delete a name from the Customer:Job list:

▌ Click **Customer Center** and click the customer name you want to delete.

▌ From the **Edit** menu, choose **Delete Customer:Job**. If the customer has been used in a transaction it will not allow you to delete it. Instead, you should make it Inactive.

▌ Click **OK**.

For more information on customers, see Chapter 7.

Vendors and Subcontractors

You can use the same process to edit your vendor and subcontractor list:

▌ Click **Vendor Center**.

▌ Click **Print** and select **Vendor List**.

To delete a name from the Vendor list:

▌ In the Vendor Center, click the vendor name you want to delete.

▌ From the **Edit** menu, choose **Delete Vendor**.

▌ Click **OK**.

For more information on vendors and subcontractors, see Chapter 8.

Other Names

This is where you enter names that aren't customers, vendors, subcontractors, or employees. Use this list for contacts you deal with occasionally. To enter names on this list:

▌ From the **Lists** menu, choose **Other Names List**.

▌ From the pull-down **Other Names** menu, choose **New**.

▌ In the New Name window, enter information as you wish.

▌ Click **OK**.

To print the list:

▌ Choose **Print List** from the **Other Names** pull-down menu.

▌ Click **Print**.

Preferences

We'll provide complete information on setting preferences in Chapter 2.

Chart of Accounts

Chapter 3 covers setting up your Chart of Accounts. To see the QuickBooks Pro Chart of Accounts list:

▌ Click **Home**, then click **Chart of Accounts**, or from the **Lists** menu, choose **Chart of Accounts**.

Items

Quicken doesn't use items, so you shouldn't have anything in your items list. For more information on setting up items, see Chapter 4. To see the QuickBooks Pro Items list:

▌ Click **Home**, then click **Items & Services**, or from the **Lists** menu, choose **Item List**.

Payroll Items

Since Quicken doesn't use payroll items, you shouldn't have anything in your payroll items list. For more information on setting up payroll items, see Chapter 5. To see the QuickBooks Pro Payroll Item list:

▌ From the **Lists** menu, choose **Payroll Item List**.

Classes

If you used classes for tracking jobs in Quicken, keep in mind that QuickBooks Pro's setup is different. To see the QuickBooks Pro Class list:

▌ From the **Lists** menu, choose **Class List**.

In QuickBooks Pro, customers and jobs go together. You'll need to create customers and jobs in QuickBooks Pro. Any jobs on the class list in Quicken will have to be associated with a customer in QuickBooks Pro. For more information on setting up jobs in QuickBooks Pro, see Chapter 7, Customers and Jobs. After you've moved the jobs to the correct spot, you can set up new classes. See Chapter 6.

Memorized Reports

If you memorized reports in Quicken, chances are those reports won't be useful in QuickBooks Pro. For more information on memorized reports, see Chapter 17, Reports.

Opening Balances

You'll want to make sure all of the balances that came over from Quicken are correct. To do this:

▌ Click **Home**, then click **Chart of Accounts**, or from the **Lists** menu, choose **Chart of Accounts**.

Verify that all of the accounts are correct. If something wasn't transferred over from Quicken to QuickBooks Pro in the conversion process, you'll need to track down what's missing and enter the transaction before you enter new transactions in QuickBooks Pro. For more information on entering transactions, see Chapters 13 through 15. For more information about QuickBooks Pro opening balances, see Chapter 10, Opening Balances.

When you've finished checking and correcting your company files, go on to Chapter 2 to set up preferences.

Section 4

Converting an Existing QuickBooks Pro 2008 Company Data File to Our Setup

To convert to our setup, you'll want to work through these chapters following our suggestions:

▌ Chapter 2 Preferences — Set up your preferences the way we suggest in Chapter 2.

▌ Chapter 3 Chart of Accounts — Compare your Chart of Accounts to ours and add, change, or inactivate accounts as needed.

▌ Chapter 4 Items — Add, change, or inactivate items as needed.

▌ Chapter 5 Payroll Items — Add, change, or inactivate payroll items as needed.

▌ Chapter 6 Classes — Add, change, or inactivate classes as needed.

▌ Chapter 7 Customers — Add, change, or inactivate customers as needed.

▌ Chapter 8 Vendors — Add, change, or inactivate vendors as needed.

▌ Chapter 9 Employees — Add, change, or inactivate employees as needed.

During this process, if you find any particular items that you want to keep but don't want to appear on any list, you can inactivate them.

To make an account, item, payroll item, class, customer and job, vendor, or employee inactive:

▌ Open the appropriate list (i.e. Lists menu, or Customer, Vendor, or Employee Center).

▌ Select the name of the item in the appropriate list.

▌ Right click the item and select **Make Inactive**.

To see everything on the list, including anything inactive:

▌ Click **Include Inactive** at the bottom of the List window.

▌ For customers, vendors and employees, click the appropriate center. In the drop-down **View** field, select **All.**

Now the item won't show up in any list. However, it's not deleted, so you can activate it again later if you wish.

Go through Chapters 10 through 16 to make sure you understand how to enter transactions correctly into QuickBooks Pro. In Chapters 17 and 18 you'll see how QuickBooks Pro will make it easier for you to get the reports you need for your business.

Chapter 2
How to Set QuickBooks Pro Preferences

Preferences are used to customize QuickBooks Pro. For instance, you can turn payroll on or off, change the accounting basis for your reports, or set an interest rate for statement finance charges. There are 19 separate windows you can use to set your preferences, with several options in each window.

In this chapter we'll cover the preferences that are specific to the needs of a construction business. Any preferences not discussed aren't critical; you can set them as you like or use the defaults. We recommend you use the preferences as they default, except as noted. In order to work with these options, you have to open a company file.

Each preference has two tabs — *My Preferences* and *Company Preferences*, as in Figure 2-1. The My Preferences tab is specific to your login name and your specific needs in QuickBooks. In a multi-user environment, you can customize QuickBooks to look and act differently for each user by customizing preferences in each My Preferences tab. The Company Preferences are universal. If you change these preferences, then the preference changes company-wide, not just for an individual user.

To set your preferences, go to the **Edit** menu, select **Preferences**, and then select, one by one, the preferences shown on the following pages.

General Preferences

QuickBooks Pro defaults to the General Preferences window when you choose the Preferences command, for the first time.

In the General Preferences window you can set defaults to move easily from one part of QuickBooks Pro to another, change the way the QuickBooks Pro windows look on your computer, and change several other items. The settings are really self-explanatory so you can just go ahead and set them.

▌ Click the defaults as you wish.

▌ Click the **Accounting** icon to move to the next set of Preferences.

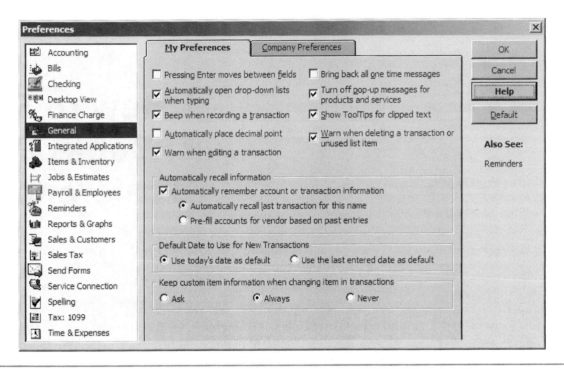

Figure 2-1
Use the My Preferences tab window to set the recommended settings for General Preferences.

If you changed the settings in the General Preferences window, QuickBooks Pro will ask if you want to save the changes. You'll be asked this question whenever you change a Preferences window and don't click **OK** to leave the window.

▌ Click **Yes** in the Save Changes window.

Accounting Preferences

Click the **Accounting** icon. Click Company Preferences tab (Figure 2-2). Here's a list of the Accounting Preferences options and what they do:

Use account numbers — Lets you use numbers for accounts in your Chart of Accounts. In Chapter 3 we'll explain why we strongly suggest you use account numbers to identify the accounts in your Chart of Accounts. For an example of what a Chart of Accounts looks like, see our sample Chart of Accounts in Chapter 3.

Show lowest subaccount only — Displays the subaccount rather than the header accounts when you view the allocation account on a transaction. This helps the data entry operator see the proper account when viewing transactions.

Require accounts — Warns and forces you to allocate each transaction to an account.

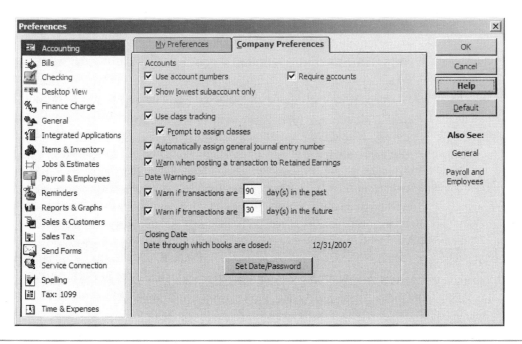

Figure 2-2
Use the Company Preferences tab window to set the recommended settings for Accounting Preferences.

Use class tracking — Lets you classify transactions into Labor, Materials, Subcontractors, Equipment Rental, and Other. This type of subtotaling is important when your tax preparer fills in the Cost of Goods section of your tax return. You'll probably also want to know exactly how much you've spent on these categories for each job.

Prompt to assign classes — We strongly suggest that you put a ✓ in the box. It will remind you to allocate each transaction to a class: Labor, Materials, Subcontractors, Equipment Rental, Other Job Related Costs or Overhead. See Chapter 6 — Classes for more information.

Automatically assign general journal entry number — Put a ✓ in the box. It's better to allow the system to automatically assign a number than to keep track of the number manually.

Warn when posting a transaction to Retained Earnings — We suggest this option be selected. It will protect your books from getting out of balance. The Retained Earning account balance is the total sum of money that has been made or lost since the beginning of the business. Each year QuickBooks automatically closes out your income and expenses and the difference gets posted into the Retained Earnings account. For example, let's say your first year in business ended with a net profit of $1,000. Your second year in business ended with a net loss of $500. The balance in Retained Earnings should be $500. Posting a transaction to Retained Earning would cause your books to get out of balance and your tax preparer to spend hours fixing the error and costing you money.

Date Warnings — Use the date warnings to be prompted if a transaction is being entered with a date more than 30 days into the future or more than 90 days in the past. The timeframes can be adjusted to whatever you wish.

Closing Date — Date through which books are closed. As soon as you submit your books to your CPA or tax preparer, you should immediately return to this screen and enter your year-end date. For example, if you've submitted paperwork or financials to your tax preparer through 12/31/08, click **Set Date/Password** and enter that date. Then enter a password in **Closing Date Password** and confirm password. Entering a date and password won't prevent data entry — it'll only warn the user that the books are closed. The user will have to enter the correct password in order to post/record a transaction in a closed year.

▌ Click the defaults in the **Accounting** Preferences window as you wish.

▌ Click the **Bills** icon to move to the next set of Preferences.

▌ Click **Yes** in the Save Changes window.

Bills Preferences

Click the **Company Preferences** tab.

If you want QuickBooks Pro to remind you about bills you need to pay, use the Bills Preferences. Initially, QuickBooks Pro uses ten days as the lead time for paying bills and putting a message in the Reminder List. See Figure 2-3 for the Bills Preferences options.

Bills are due — Typically, this is 30 days. Enter a different number of days if you wish.

Warn about duplicate bill numbers from same vendor — Select this if you want to be notified if you try to enter the same bill twice. This is a nice feature if you get lots of bills from certain vendors. It will help you avoid over-paying a vendor, or paying twice for the same bill.

Automatically use discounts and credits — Select this if any of your vendors let you take a discount if you pay early (or within 10 days). Enter a default account number to record all the discounts you receive. We used account number 7800 Trade Discount.

▌ Click the **Checking** icon to move to the next set of Preferences.

▌ Click **Yes** in the Save Changes window.

Checking Preferences

The My Preferences tab of the Checking Preferences window will help you automatically link bank accounts to specific functions or tasks. For example, let's suppose you have a payroll checking account, a general checking account, and a savings account. If you usually pay payroll from the payroll

Figure 2-3

Use the Company Preferences tab window to set the recommended settings for Bills Preferences.

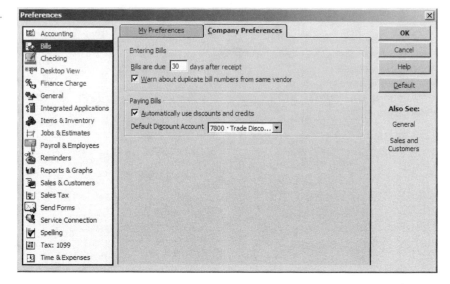

Figure 2-4

Use the My Preferences tab of Checking Preferences to set preferences for a certain person or computer.

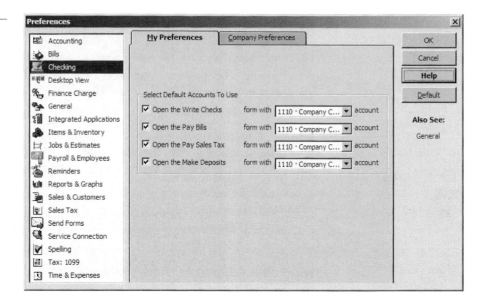

account, write checks from the general checking account, and deposit money into the savings account, you can use the My Preferences tab of the Checking Preferences window to link the appropriate bank accounts whenever you perform one of these actions. Having the correct bank account appear automatically should cut down on data entry errors. However, keep in mind that the My Preferences tab in this window lets you set preferences for a certain person or a certain computer. If you sign on with a different name, you'll need to set the preferences again for that sign-on name. You can see the setup we used in Figure 2-4.

The Company Preferences tab of the Checking Preferences window lets you set up universal preferences for printing and processing checks.

Figure 2-5

Use the Company Preferences tab of Checking Preferences to set general preferences for printing and processing checks.

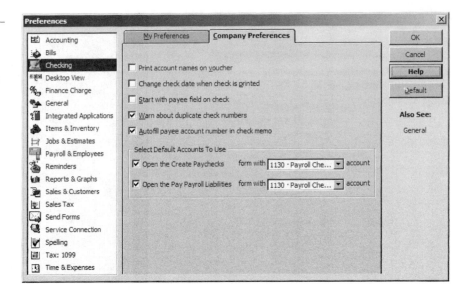

Here's what the options shown in Figure 2-5 do:

Print account names on voucher — If you use checks with voucher attachments, the payee, date, account name, memo, amount, and total amount will appear on the voucher. If you don't select this, QuickBooks Pro omits the account name(s) but prints the rest of the information.

Change check date when check is printed — Shows the date the check was printed, not the date it was created in QuickBooks Pro. Check this if you enter your checks on a different day than you print or send them out.

Start with payee field on check — Lets you skip check number and date fields and enter information in the payee field. QuickBooks Pro will fill in the date and next check number automatically. Select this option if you're current with entering your data and stick to a consistent check number sequence.

Warn about duplicate check numbers — It's a good idea to select this to reduce the chance of duplicating a check entry.

Autofill payee account number in check memo — If you select this box, the vendor account will print on the memo field on the check.

Select Default Accounts To Use — Enter the correct default accounts you use for payroll here. If you have more than 20 employees, we think it's a good idea to use a separate payroll checking account. We've found that it can take too long to reconcile a general checking account if you process payroll each week and print more than 20 checks for each payroll.

▌ Click the defaults and enter accounts as you wish in the **My Preferences** and **Company Preferences** windows.

▌ Click the **Finance Charge** icon to move to the next set of Preferences.

▌ Click **Yes** in the Save Changes window.

Figure 2-6

Use the Company Preferences tab window to set the recommended settings for Finance Charge Preferences.

Finance Charge Preferences

Now you're ready to set the Finance Charge Preferences. See Figure 2-6.

It's up to you whether or not to charge your customers a fee for late payments. Many companies use the finance charge as a threat to keep payments timely, but in practice few of them actually collect the charges. If you decide to charge for late payments, we suggest you explain the charges very clearly in all your contract documents so your customers aren't surprised when they're hit with these "extras."

Finance Charge Preferences let you set the percentage and minimum finance charge you want to add to overdue invoices, and the grace period. In Figure 2-6 it's set at 10 days. When you enter an amount in Minimum Finance Charge, the amount is applied regardless of the amount overdue.

Here's another example of a grace period. If your terms are net 30 and you want to start charging interest after a 30-day grace period, enter 30. Then if you send out an invoice dated January 1, QuickBooks Pro waits until January 31 to assess finance charges on any unpaid amount.

If you use finance charges, you must enter the account you use to track them. If you don't have this account in your Chart of Accounts, you need to create it now.

▌ At **Finance Charge Account**, pull down the menu and choose **Add New** (at the top of the list). This will bring up the New Account window. Figure 2-7 shows how we entered account 7010 Interest Income (You don't need to add this account if you started with the Company file from the CD).

▌ In **Account Type**, choose **Other Income** from the pull-down list.

Figure 2-7
Use the New Account window to create an account that will track your finance charges.

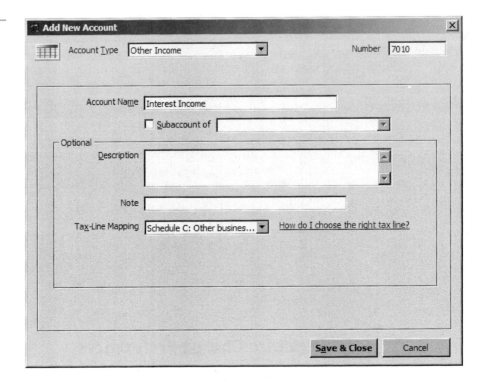

- In **Number**, we entered **7010**. That's Interest Income from the Chart of Accounts (see Chapter 3).

- In **Account Name**, we entered **Interest Income**.

- In **Tax Line Mapping**, choose **Schedule C: Other business income** from the pull-down list.

- Click **Save & Close**.

Unless you want to play hardball, we don't suggest you select Assess finance charges on overdue finance charges. See Figure 2-6.

- Click the **Items & Inventory** icon to move to the next set of Preferences.

- Click **Yes** in the Save Changes window.

Items & Inventory Preferences

If your business is inventory-based, you can select your preferences for inventory and purchase orders in the Items & Inventory Preferences window. See Figure 2-8 for the Items and Inventory options.

Inventory and purchase orders are active — Select this if you plan to use purchase orders to track which materials are on order, when you ordered them, who you ordered them from, or if you plan to use purchase orders to track committed costs (see Chapter 14).

Figure 2-8

Use the Company Preferences tab window to set the recommended settings for Items & Inventory Preferences.

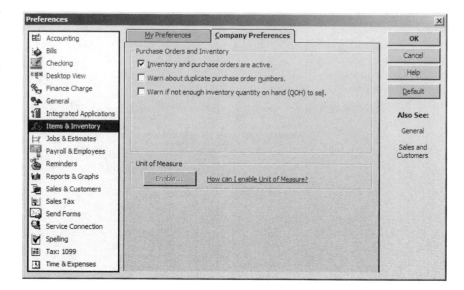

Warn about duplicate purchase order numbers — Select this if you want to be warned when you create a duplicate purchase order.

▌ Click on the **Jobs & Estimates** icon to move to the next set of Preferences.

▌ Click **Yes** in the Save Changes window.

Jobs & Estimates Preferences

Whenever you click Customer Center, you get a list of your customers, their jobs, and one of five possible descriptions for the status of each job. The default descriptions are Pending, Awarded, In progress, Closed, and Not awarded. If you want to use other descriptions for job status, enter them on the Jobs and Estimates Preferences screen. To change a description, just enter the new description. See Figure 2-9.

▌ Set remaining defaults as you prefer. If you will be billing customers based on a percent complete by phase of construction, answer **Yes** to the question **Do You Do Progress Invoicing?** For more information on this, see Chapter 13.

▌ Click the **Payroll & Employees** icon to move to the next set of Preferences.

▌ If you get the Save Changes window, click **Yes**.

Figure 2-9
Use the Company
Preferences tab window
to set the recommended
settings for the Jobs &
Estimates Preferences.

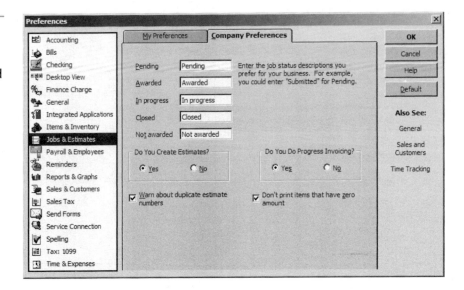

Figure 2-10
Use the Company
Preferences tab window
to set the recommended
settings for
Payroll & Employees
Preferences.

Payroll & Employees Preferences

Payroll and Employee Preferences are extremely important. The way you set these preferences determines how QuickBooks Pro will associate costs, such as payroll taxes and wages, to job cost reports. See Figure 2-10 for the settings we recommend.

Full payroll — Uses the QuickBooks Pro payroll module to allocate gross wages and payroll taxes to jobs. You'll probably want to select this even if you use a payroll service. For a more detailed discussion, see Chapter 15 on payroll.

Figure 2-11

Use the Employee Defaults window to set the defaults you'll use for each new employee you enter into QuickBooks Pro.

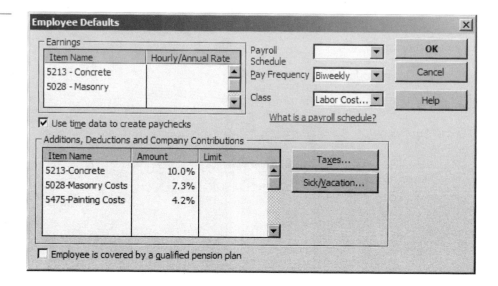

Job Costing, Class and Item tracking for paycheck expenses — Make sure you select this preference. It will automatically disburse employer payroll taxes to the job, item (job phase), and class (labor).

Assign one class per — Since we'll use one class (labor) for all transactions on a timecard, we'll select **Entire paycheck** to assign the labor class to the entire timecard. This will save you time when you enter timecards and process payroll.

You can also set the defaults you want to use when entering employee information by clicking the **Employee Defaults** button. Let's do that now to display the window shown in Figure 2-11.

▌ Click **Employee Defaults**.

The defaults you set here will appear on the New Employee, Payroll Info Tab window each time you enter a new employee.

▌ Change the template defaults as you wish.

▌ Click **OK** to save the template.

If you are using QuickBooks Enhanced Payroll service, you will have access to a feature that helps you prepare your workers' compensation report. Although it doesn't print the report for you, or allocate workers' comp costs to jobs, it will give you the information you'll need to fill out the report. The reason QuickBooks doesn't prepare your workers' comp report for you is because there is a vast array of different insurance companies you can purchase insurance from and each one has a different report type. To use the new feature, click **Set Preferences** in the Workers' Compensation box.

Figure 2-12

Use the Company
Preferences tab window
to set the recommended
settings for
Reminders Preferences.

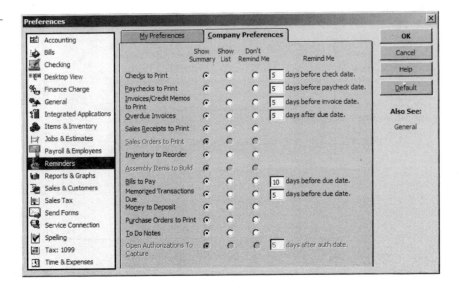

Some changes you make to Payroll & Employees Preferences may cause QuickBooks Pro to advise you of the effects of the changes you've made. You can just select **OK** for these warnings.

❚ Click the **Reminders** icon to move to the next set of Preferences.

❚ Click **Yes** in the Save Changes window.

Reminders Preferences

QuickBooks Pro has an extensive set of task reminders. It can remind you to print checks, payroll checks, invoices, or to pay bills and deposit money. These can be handy functions, especially if you have several people working on your accounting system. Or they can be annoyances if you already have a schedule and live by it. QuickBooks Pro lets you "have it your way" by turning these Reminder Preferences on or off.

The Reminders Preferences screen gives you three options for each task — Show Summary, Show List, or Don't Remind Me. For some of the tasks you can set the number of days before you're warned. To follow our example, set your Reminders Preferences to match Figure 2-12 for now.

❚ Click the **Reports & Graphs** icon to move to the next set of Preferences.

❚ Click **Yes** in the Save Changes window.

Figure 2-13

This Company Preferences tab window shows the recommended settings for Reports & Graphs.

Reports and Graphs Preferences

If you select Accrual in the Summary Reports Basis section of the Company Preferences tab window of Reports and Graphs Preferences, QuickBooks Pro will include unpaid vendor bills and customer invoices when figuring income and expenses, Accounts Payable, and Accounts Receivable, even though you haven't received the money. If you select Cash, only the vendor bills and customer invoices that have been paid will show up on reports.

Generally, it's a good idea to run your business on an accrual basis so you can see how much money you've earned and committed to spend over a given time period. The cash basis shows only how much money you received in a month and how much money you paid out. The accrual basis will give you a better picture of monthly income versus expenses.

For now, select the options as shown in Figures 2-13 and 2-14. You can do any fine-tuning later when you have more experience using reports.

▌ Click the **Sales & Customers** icon to move to the next set of Preferences.

▌ Click **Yes** in the Save Changes window.

Sales and Customers Preferences

Many options on Sales & Customers Preferences aren't important to the construction industry, so we don't describe them here. There are three important preferences to select and we've explained below how you should make the right selection.

Figure 2-14

Use the My Preferences tab window to set your own preferences for Reports & Graphs.

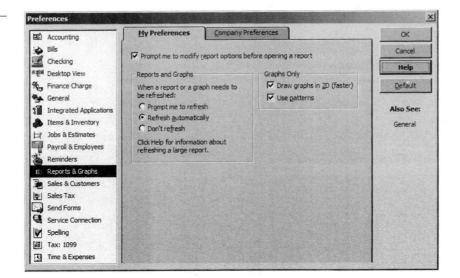

Select **Warn about duplicate invoice numbers** if you want to track invoices by number and want to avoid issuing two invoices with the same number.

We recommend that you DO NOT select any options in the Receive payments boxes.

▌ Click the **Company Preferences** tab.

▌ Select items as shown in Figure 2-15.

▌ Click the **Sales Tax** icon to move to the next set of Preferences.

▌ Click **Yes** in the Save Changes window.

Sales Tax Preferences

These Preferences are self-explanatory and will depend on the requirements in your state. See Figure 2-16.

▌ Select items as you wish.

▌ Click the **Send Forms** icon to move to the next set of Preferences.

▌ Click **Yes** in the Save Changes window.

Send Forms Preferences

If you choose to send forms, such as a customer invoice, via email, you can set the Send Forms preferences to automatically attach a cover message. See Figure 2-17.

▌ Select items as you wish.

▌ Click the **Service Connection** icon to move to the next set of Preferences.

▌ Click **Yes** in the Save Changes window.

Figure 2-15

Use the Company Preferences tab window to set the recommended settings for the Sales & Customers Preferences.

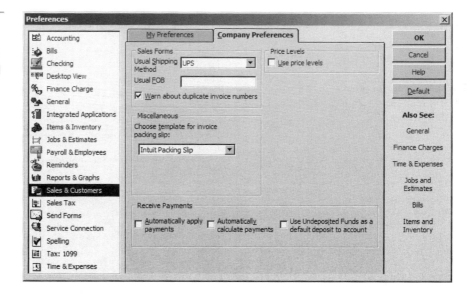

Figure 2-16

Use the Company Preferences tab window of Sales Tax Preferences to specify your company preferences for sales tax.

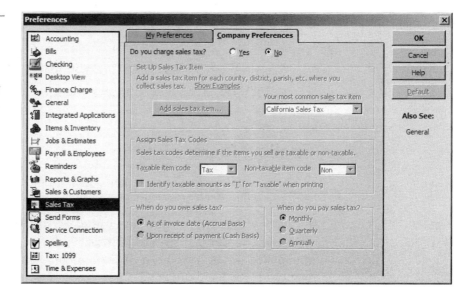

Figure 2-17

Use the Company Preferences tab window of Send Forms Preferences to set your company preferences for sending forms.

Service Connection Preferences

Service Connection Preferences lets you choose how you want to handle connections to QuickBooks services. If you're just getting started with QuickBooks, you're probably not using any of these services. In this case, leave the settings as they are and go to the next set of Preferences. See Figure 2-18.

▌ Select the items you want.

▌ Click the **Spelling** icon to move to the next set of Preferences.

▌ Click **Yes** in the Save Changes window.

Spelling Preferences

With the Spelling Preferences you can send invoices to customers without the worry that you've misspelled something. See Figure 2-19.

We recommend that you select **Always check spelling before printing, saving, or sending supported forms.** Then QuickBooks Pro will find any misspelled words so you can fix them before your customer finds them and begins to wonder about you.

▌ Select items as you wish.

▌ Click the **Tax:1099** icon to move to the next set of Preferences.

▌ Click **Yes** in the Save Changes window.

Figure 2-18
Service Connection preferences determine how you connect to Internet applications.

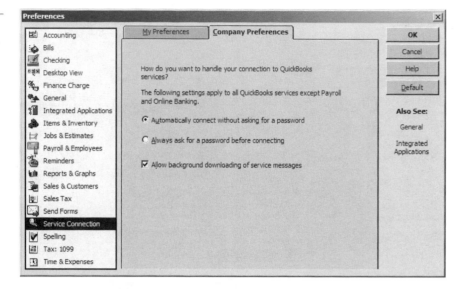

Figure 2-19
Use the My Preferences tab window of the Spelling Preferences to get QuickBooks Pro to check the spelling on your forms.

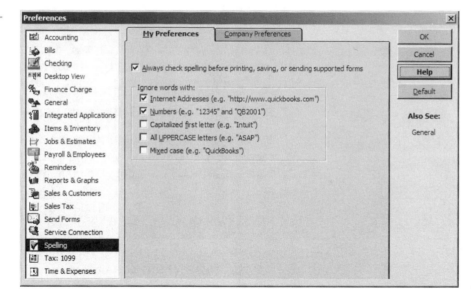

Tax: 1099 Preferences

1099s are forms that your business files with the IRS to report the total amount your company paid each vendor or subcontractor for the year starting January 1st and ending December 31st. You must file a 1099 for each vendor or subcontractor who has provided over $600 of labor during the year. If you work with vendors or subcontractors to whom you send 1099 forms, you can set up QuickBooks Pro to track all 1099-related payments. At the end of the year, QuickBooks can print your 1099 forms. You're not required to send 1099 forms to material suppliers, or to vendors from whom you purchase materials.

Figure 2-20

Use the Company
Preferences tab window
to specify your company
preferences for 1099 tax
categories.

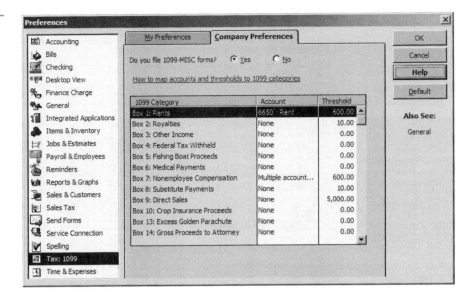

It's easier to complete this section if you've already set up your Chart
of Accounts. So, you may want to do that first and then come back here. If
you're ready now, proceed and see Figure 2-20.

▌ First answer the **Do you file 1099-MISC forms?** question.

If the answer is yes, you need to link the appropriate accounts with
each 1099 category. Look in the list of 1099 categories for the categories in which you report amounts to the IRS. Most construction
businesses report amounts in Box 7: Nonemployee compensation.

▌ Click in the **Account** column next to Box 7: Nonemployee Compensation. In the drop-down list that appears, choose the account you'll
use to track 1099 payments to your vendors. If you use more than one
account, click **Multiple accounts** at the top of the account list and then
select each individual account, as shown in Figures 2-21 and 2-22.

Note that each 1099 category must have its own unique account. For
example, if you report Nonemployee Compensation (Box 7), and you have
a Cost of Goods Sold account named Job Related Costs, you can select it
as the account for tracking Nonemployee Compensation. But once you
select Job Related Costs for Box 7, you can't use it to track any of the other
1099 categories.

▌ Enter items as you wish.

▌ Click the **Time & Expenses** icon to move to the next set of
Preferences.

▌ Click **Yes** in the Save Changes window.

Figure 2-21

You need to select the checked accounts shown here, and in Figure 2-22.

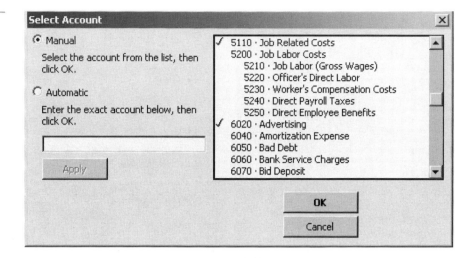

Figure 2-22

Select all accounts that you might allocate a sub-contractor or outside consultant expense to, so a 1099 can be created at year end.

Time & Expenses Preferences

It's very easy to track time spent on a job using the time tracking function in QuickBooks Pro. This is one of the most important functions QuickBooks Pro can perform for you. Accurate time tracking lets you compare your actual labor to your estimated labor. In other words, it's a link in the "feed-back" loop from estimate to job cost and back to estimate.

In **First Day of the Work Week** select the day your printed timecards start. In our example, we start our work week on Monday and end the week on Sunday. If you start your week on Thursday and end on Wednesday, select Thursday. If you pay your employees biweekly or on the 1st and 15th, select the day you've been using on your timecards. When

Figure 2-23

Use the Company
Preferences tab window
to set your Time &
Expenses Preferences.

QuickBooks processes payroll it will know to take the dates that make up
the pay period, not the full week entered.

▌ Select items shown in Figure 2-23, as you wish.

▌ Click **OK**.

We recommend that you don't select **Track reimbursed expenses
as income**. That way you'll get reports showing customer invoices as
construction income and vendor bills as job costs. If you do select it,
QuickBooks Pro records income and expense on time and materials jobs by
shifting amounts into, and out of, the same account.

Now you should have your QuickBooks Pro company file set up and
your preferences customized. The next step is to create or change your
Chart of Accounts. In the next chapter we'll explain what a Chart of
Accounts is and how to add, delete, or edit accounts.

Chart of Accounts

The Chart of Accounts is the backbone of your accounting system. That's why it's so important that you understand how it works. You can think of the Chart of Accounts as a file cabinet with a file for each type of accounting information you want to track. For example, if you need to know how much money you spend on postage, you'll set up a file (an account in the Chart of Accounts) for Postage Expense.

The Chart of Accounts

Although you aren't required to use account numbers in your Chart of Accounts in QuickBooks Pro, we suggest that you do. Here are the standard Chart of Accounts numbers:

1000 - 1999 Assets
2000 - 2999 Liabilities
3000 - 3999 Capital
4000 - 4999 Income or Revenue
5000 - 5999 Job Costs
6000 - 6999 Overhead Costs
7000 - 7999 Other Income
8000 - 8999 Other Expense

Now let's look at each item on this list.

Assets

Assets are things your company owns. They're usually divided into two groups — current assets and fixed assets. Current assets are generally numbered from 1000 - 1499. These are assets that you can easily turn into cash, such as checking accounts, savings accounts, money market and CD accounts, Accounts Receivable, and inventory. So you might want to use account number 1100 for your company checking account because a checking account is a current asset.

Fixed assets are usually numbered from 1500 - 1999. These are items you would have to sell to generate cash. Automobiles, equipment, and land are examples of fixed assets. Let's look at an example. Last year Twice Right Construction bought a new table saw for $1,100. Since the saw cost over $500 (which is the minimum amount considered a fixed asset), the purchase was entered to an asset account rather than to an expense account.

Liabilities

Liabilities are funds your company owes. For example, last year Twice Right Construction borrowed $20,000 from We Trust U Bank. When the $20,000 loan was deposited to the checking account, the deposit was entered in the liability account Bank Loans, not an income account such as Construction Revenue.

Capital

Your capital account structure depends on whether your company is organized as a sole proprietorship, partnership, or corporation.

If your company is a *sole proprietorship*, you need a Capital account and an Owner's Drawing account. Use the Capital account to keep track of the total amount of money you've invested since starting the business, plus or minus the net profit or loss each year since you started the business. Use the Owner's Drawing account for money you take out of the business for personal use, such as checks to the grocery store, dry cleaners, ATM transactions, and your salary. It's important to keep in mind that the owner of a sole proprietorship doesn't get a regular "employee" paycheck with money deducted for payroll taxes. Instead you pay quarterly estimated taxes, which you should always allocate to the Owner's Drawing account.

If your company is a *partnership* or *LLP (Limited Liability Partnership)*, you need to set up Capital and Drawing accounts for each partner.

If your company is an *S* or *C corporation* or an *LLC*, it has a Common Stock account and sometimes a Preferred Stock account. Common stock and preferred stock represent the total sum of stock the company has issued.

Income or Revenue

Income or revenue is the income you get from the everyday way you do business, such as remodeling income or construction revenue. You would put other income or revenue, such as rents from buildings you own, in a separate Other Income account with a number between 7000 - 7999.

Job Costs

Job Costs (also called Cost of Goods Sold in the manufacturing industry) are all the costs of building your product. If you're a home builder, job costs are whatever it costs you to build a home, including the direct labor, materials, subcontractors, dump fees, and equipment rental. If you design homes, job costs include all the costs associated with designing a home, such as design labor, drafting materials, supplies, and engineering costs. If you do both designing and building, you'll have both sets of costs. It's important to note here that the *sample.qbw* and *company.qbw* data files on the CD-ROM included with this book are set up to have just *one* total Job Cost. We use classes to break down that total into labor, materials, subcontractors, equipment rental, and any other costs related to a particular job.

Overhead Costs

Overhead Costs are fixed costs you have even if you run out of work. Rent, telephone, insurance, and utilities are overhead costs.

Other Income

Other Income is income you earn outside the normal way you do business, including interest income, and gain on the sale of an asset, insurance settlement, or stock sale.

Other Expense

Other Expense is an expense that's outside of your normal business, such as a loss on the sale of an asset or stockbroker fees.

How to Use the Sample Chart of Accounts

If you haven't already done so, you may want to take a look at the Chart of Accounts in the *sample.qbw* file on the CD-ROM you got with this book. It has example transactions, a few jobs, vendors, customers, items, classes, memorized reports, memorized transactions, and a Chart of Accounts. You can use the sample company to test a new idea and get acquainted with QuickBooks Pro.

If you're ready for the real thing, you can use the Chart of Accounts in the *company.qbw* file on the CD-ROM for a jump start. This is a general Chart of Accounts for a sole proprietor building contractor. We've added other accounts we think you need for building contractor work. See the Chart of Accounts listings on the next pages for examples specific to a sole proprietor or partnership, and a corporation.

Chart of Accounts
Sole Proprietor or Partnership

Account	Type
1110 · Company Checking Account	Bank
1111 · Adjustment Register	Bank
1120 · Company Savings Account	Bank
1130 · Payroll Checking Account	Bank
1140 · Petty Cash Account	Bank
1210 · Accounts Receivable	Accounts Receivable
1300 · Inventory Asset	Other Current Asset
1310 · Employee Advances	Other Current Asset
1320 · Retentions Receivable	Other Current Asset
1330 · Security Deposit	Other Current Asset
1340 · Vendor Deposits	Other Current Asset
*1390 · Undeposited Funds	Other Current Asset
1400 · Refundable Workers' Comp Deposit	Other Current Asset
1510 · Automobiles & Trucks	Fixed Asset
1520 · Computer & Office Equipment	Fixed Asset
1530 · Machinery & Equipment	Fixed Asset
1540 · Accumulated Depreciation	Fixed Asset
2010 · Accounts Payable	Accounts Payable
2050 · MasterCard Payable	Credit Card
2060 · Visa Card Payable	Credit Card
2100 · Payroll Liabilities	Other Current Liability
2200 · Customer Deposits	Other Current Liability
2201 · Sales Tax Payable	Other Current Liability
2240 · Workers' Comp Payable	Other Current Liability
2300 · Loans Payable	Other Current Liability
2310 · Line of Credit	Other Current Liability
2410 · Construction Loan	Other Current Liability
2510 · Truck Loan	Long Term Liability
3000 · Opening Balance Equity	Equity
**3100 · Owner's Equity	Equity
3110 · Owner's Capital	Equity
3130 · Owner's Draws	Equity
3910 · Retained Earnings	Equity
3999 · Owner's Time to Jobs	Equity
4110 · Construction Income	Income
4810 · Vendor Refunds	Income
4910 · Workers' Comp Dividend	Income
5110 · Job Related Costs	Cost of Goods Sold
5200 · Job Labor Costs	Cost of Goods Sold
5210 · Job Labor (Gross Wages)	Cost of Goods Sold
5220 · Officer's Direct Labor	Cost of Goods Sold
5230 · Workers' Comp Costs	Cost of Goods Sold
5240 · Direct Payroll Taxes	Cost of Goods Sold
5250 · Direct Employee Benefits	Cost of Goods Sold
6020 · Advertising	Expense

6040 · Amortization Expense	Expense
6050 · Bad Debt	Expense
6060 · Bank Service Charges	Expense
6070 · Bid Deposit	Expense
6075 · Bond Expense	Expense
6090 · Business License & Fees	Expense
6100 · Car/Truck Expense	Expense
6101 · Gas & Oil	Expense
6103 · Repairs & Maintenance	Expense
6105 · Registration & License	Expense
6107 · Insurance-Auto	Expense
6130 · Cleaning/Janitorial	Expense
6135 · Computer Supplies/Equipment	Expense
6140 · Contributions	Expense
6150 · Depreciation Expense	Expense
6160 · Dues and Subscriptions	Expense
6180 · Insurance	Expense
6181 · Disability Insurance	Expense
6182 · Liability Insurance	Expense
6185 · Workers' Comp	Expense
6200 · Interest Expense	Expense
6201 · Finance Charge	Expense
6202 · Loan Interest	Expense
6203 · Credit Card Interest	Expense
6300 · Office Expense	Expense
6330 · Office Supplies	Expense
6350 · Office Equipment (under $500)	Expense
6500 · Payroll Expenses (office)	Expense
6501 · Payroll (office staff)	Expense
6502 · Payroll Tax Expense	Expense
6503 · Officer's Wages	Expense
6504 · Designer's Wages	Expense
6508 · Vac/Holiday/Sick Pay	Expense
6509 · Employee Bonus	Expense
6510 · Employee Benefits	Expense
6570 · Professional Fees	Expense
6571 · Accounting	Expense
6572 · Legal Fees	Expense
6573 · Computer Consultants	Expense
6610 · Postage and Delivery	Expense
6650 · Rent	Expense
6670 · Repairs	Expense
6671 · Building Repairs	Expense
6672 · Computer Repairs	Expense
6673 · Equipment Repairs	Expense
6800 · Telephone	Expense
6820 · Taxes	Expense
6830 · Training & Conferences	Expense
6900 · Travel & Entertainment	Expense
6901 · Entertaining Clients	Expense

6902 · Meals	Expense
6903 · Air Fare	Expense
6904 · Hotels/Lodging	Expense
6920 · Tools & Machinery (under $500)	Expense
6970 · Utilities	Expense
7010 · Interest Income	Other Income
7030 · Other Income	Other Income
7800 · Trade Discounts	Other Income
8010 · Other Expenses	Other Expense
*2 · Purchase Orders	Non-Posting
*4 · Estimates	Non-Posting

*QuickBooks Pro sets up these accounts for you automatically. The account number for undeposited funds may not be the same number you listed.

** If your business is a partnership between Partner 1 and Partner 2, this section would look like this:

3100 · Partner 1 Equity	
3110 · Partner 1 Investments	Equity
3120 · Partner 1 Drawing Account	Equity
3200 · Partner 2 Equity	Equity
3210 · Partner 2 Investments	Equity
3220 · Partner 2 Drawing Account	Equity

Chart of Accounts
Corporation

Account	Type
1110 · Company Checking Account	Bank
1111 · Adjustment Register	Bank
1120 · Company Savings Account	Bank
1130 · Payroll Checking Account	Bank
1140 · Petty Cash Account	Bank
1210 · Accounts Receivable	Accounts Receivable
1300 · Inventory Asset	Other Current Asset
1310 · Employee Advances	Other Current Asset
1320 · Retentions Receivable	Other Current Asset
1330 · Security Deposit	Other Current Asset
1340 · Vendor Deposits	Other Current Asset
*1390 · Undeposited Funds	Other Current Asset
1400 · Refundable Workers' Comp Deposit	Other Current Asset
1510 · Automobiles & Trucks	Fixed Asset
1520 · Computer & Office Equipment	Fixed Asset

1530 · Machinery & Equipment	Fixed Asset
1540 · Accumulated Depreciation	Fixed Asset
2010 · Accounts Payable	Accounts Payable
2050 · MasterCard Payable	Credit Card
2060 · Visa Card Payable	Credit Card
2100 · Payroll Liabilities	Other Current Liability
2200 · Customer Deposits	Other Current Liability
2201 · Sales Tax Payable	Other Current Liability
2240 · Workers' Comp Payable	Other Current Liability
2300 · Loans Payable	Other Current Liability
2310 · Line of Credit	Other Current Liability
2410 · Construction Loan	Other Current Liability
2510 · Truck Loan	Long Term Liability
3000 · Opening Balance Equity	Equity
3100 · Common Stock	Equity
3200 · Shareholder Distribution	Equity
3910 · Retained Earnings	Equity
4110 · Construction Income	Income
4810 · Vendor Refunds	Income
4910 · Workers' Comp Dividend	Income
5110 · Job Related Costs	Cost of Goods Sold
5200 · Job Labor Costs	Cost of Goods Sold
5210 · Job Labor (Gross Wages)	Cost of Goods Sold
5220 · Officer's Direct Labor	Cost of Goods Sold
5230 · Workers' Comp Costs	Cost of Goods Sold
5240 · Direct Payroll Taxes	Cost of Goods Sold
5250 · Direct Employee Benefits	Cost of Goods Sold
6020 · Advertising	Expense
6040 · Amortization Expense	Expense
6050 · Bad Debt	Expense
6060 · Bank Service Charges	Expense
6070 · Bid Deposit	Expense
6075 · Bond Expense	Expense
6090 · Business License & Fees	Expense
6100 · Car/Truck Expense	Expense
6101 · Gas & Oil	Expense
6103 · Repairs & Maintenance	Expense
6105 · Registration & License	Expense
6107 · Insurance-Auto	Expense
6130 · Cleaning/Janitorial	Expense
6135 · Computer Supplies/Equipment	Expense
6140 · Contributions	Expense
6150 · Depreciation Expense	Expense
6160 · Dues and Subscriptions	Expense
6180 · Insurance	Expense
6181 · Disability Insurance	Expense
6182 · Liability Insurance	Expense
6185 · Workers' Comp	Expense
6200 · Interest Expense	Expense
6201 · Finance Charge	Expense

6202 · Loan Interest	Expense
6203 · Credit Card Interest	Expense
6300 · Office Expense	Expense
6330 · Office Supplies	Expense
6350 · Office Equipment (under $500)	Expense
6500 · Payroll Expenses (office)	Expense
6501 · Payroll (office staff)	Expense
6502 · Payroll Tax Expense	Expense
6503 · Officer's Wages	Expense
6504 · Designer's Wages	Expense
6508 · Vac/Holiday/Sick Pay	Expense
6509 · Employee Bonus	Expense
6510 · Employee Benefits	Expense
6570 · Professional Fees	Expense
6571 · Accounting	Expense
6572 · Legal Fees	Expense
6573 · Computer Consultants	Expense
6610 · Postage and Delivery	Expense
6650 · Rent	Expense
6670 · Repairs	Expense
6671 · Building Repairs	Expense
6672 · Computer Repairs	Expense
6673 · Equipment Repairs	Expense
6800 · Telephone	Expense
6820 · Taxes	Expense
6830 · Training & Conferences	Expense
6900 · Travel & Entertainment	Expense
6901 · Entertaining Clients	Expense
6902 · Meals	Expense
6903 · Air Fare	Expense
6904 · Hotels/Lodging	Expense
6920 · Tools & Machinery (under $500)	Expense
6970 · Utilities	Expense
7010 · Interest Income	Other Income
7030 · Other Income	Other Income
7800 · Trade Discounts	Other Income
8010 · Other Expenses	Other Expense
*2 · Purchase Orders	Non-Posting
*4 · Estimates	Non-Posting

*QuickBooks Pro sets up these accounts for you automatically. The account number for undeposited funds may not be the same number you listed.

Figure 3-1

You can change the type, number, name, description, note, tax-related information, or opening balance for any existing account in the Edit Account window.

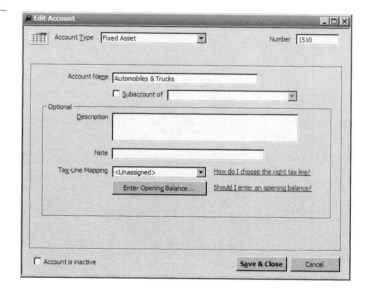

Take a little time now to compare the account list on the last 5 pages to your existing Chart of Accounts. You'll probably need to make changes for your business. For example, if you currently use expense accounts that you don't find listed in our Chart of Accounts, you'll need to add those accounts. But we'll add a word of caution here. Keep in mind that items such as payroll items and material items are linked to 5110 Job Related Costs and 4110 Construction Income accounts in the Chart of Accounts. You don't want to change these two account numbers or names.

Change, Add to, and Print Your Chart of Accounts

To change an existing account:

❚ Click **Home**, then click the **Chart of Accounts** icon, or from the **Lists** menu, choose **Chart of Accounts**.

❚ Click once to highlight the account you want to edit. Pull down the **Account** menu at the bottom of the window and choose **Edit Account**.

Figure 3-1 shows an example of the screen you'll see.

❚ Change the information for the account and click **Save & Close**. For information on Opening Balances see Chapter 10.

After you change an account, you should make sure all transactions are correctly assigned. To do that:

❚ Click **Home**, then click the **Chart of Accounts** icon, or from the **Lists** menu, choose **Chart of Accounts** and select the account you changed.

❚ At the bottom of the Chart of Accounts window, pull down the **Activities** menu and choose **Use Register**.

Figure 3-2
Use the New Account window to create a new account in your Chart of Accounts.

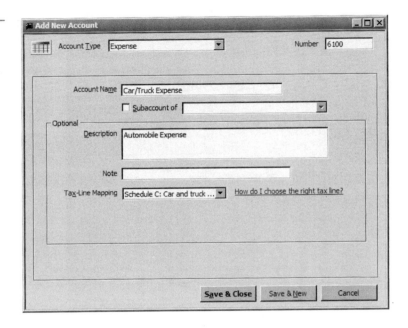

Now check the register and make sure everything listed really should be assigned to the account. If a transaction belongs in another account, double click on the transaction in the register and reassign it to the correct account.

To add a new account to your Chart of Accounts:

▌ Click **Home**, then click **Chart of Accounts**, or from the **Lists** menu, choose **Chart of Accounts**.

▌ Pull down the **Account** menu at the bottom of the window and choose **New**.

▌ Select the account type. Click **Continue**.

▌ Add the information for the account and click **Save & Close**.

Figure 3-2 shows an example of how to enter a new account for car/truck expenses to your Chart of Accounts.

To get more detail about a specific account, you can break it down by subaccounts. Figure 3-3 shows an example of a new subaccount window for subaccount 6101 Gas & Oil.

▌ Add the information for the subaccount and click **Save & Close**.

After you've finished changing your Chart of Accounts, you can print a new list and keep it handy for future reference. To do that:

▌ From the pull-down **Reports** menu in the Chart of Accounts window, choose **Account Listing**.

▌ Click **OK** to display the report.

▌ Click **Print**.

Figure 3-3

You can create subaccounts to get a more detailed breakdown of an account.

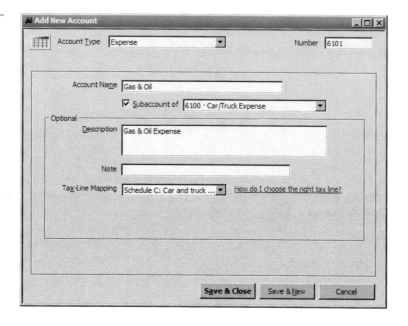

QuickBooks Premier Contractor Edition Features

If you're using the QuickBooks Premier Contractor (or Accountant) Edition, you have access to a few additional features, including:

- Fixed Asset Tracking
- Loan Manager
- Cash Flow Projector
- Forecasting
- Business Planning
- Expert Analysis

Fixed Asset Tracking

Fixed asset tracking allows you to record each asset you purchase in the Fixed Asset Item List located under the Lists menu. Use this list to track each item you purchase, including: property, automobiles and trucks, equipment, large tools, and computer purchases. We suggest you "write off," or expense, items that cost under $500. In other words, only use the Fixed Asset Item List for items that cost over $500. See Figure 3-4.

Unfortunately, the Fixed Asset Item List doesn't set up depreciation schedules or post depreciate for you. Instead, it gives you one convenient location to store information about an asset, such as date of purchase, purchase price, where you bought it, when, and for how much you sold the

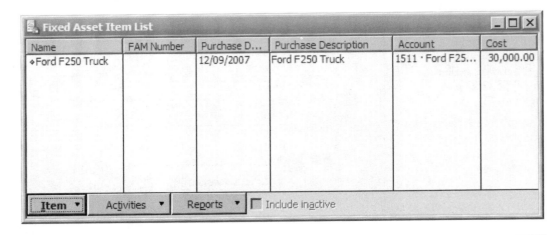

Figure 3-4
The Fixed Asset Item List is only used for items over $500.

asset, and so on. Your accountant can use the information from the fixed asset item to calculate depreciation, but you or your accountant will have to post a general journal entry to record depreciation.

Loan Manager

The Loan Manager is a payment calculator that computes the principal and interest portion of your loan payment, plus handles any fees or charges. It helps you track loans based on the information in your Long Term Liability and Other Current Liability accounts in QuickBooks. When you use the Loan Manager, you can track all of your loans in one location and be reminded of upcoming payments. Use the Loan Manager to:

- Add and remove loans you want to track.
- View payment schedules.
- Set up loan payments.
- Analyze different loan scenarios.

The Loan Manager creates payment schedules that you can view and print, allowing you to track loan-related information on a per-payment and per-total-payments basis. Plus, when you need to edit or make changes to a loan, the Loan Manager recalculates your payment information and payment schedule. Before using the Loan Manager, walk through our example below.

To set up a loan in the Loan Manager:

1. First, make sure you have a liability account set up for the loan. For example, if you took out a loan to purchase a truck, set up a loan account for the truck. See Figure 3-5.

Figure 3-5

Use the Add New Account window to create aliability account for a loan.

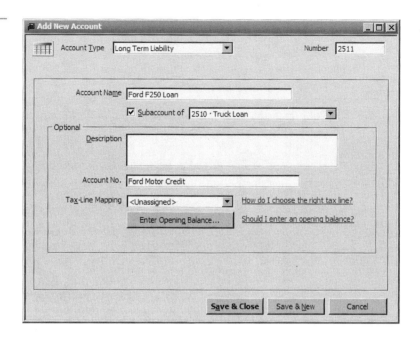

- ▌ Click **Home**, then click **Chart of Accounts**, or from the **Lists** menu, choose **Chart of Accounts**.

- ▌ Pull down the **Account** menu at the bottom of the window and choose **New**.

- ▌ Select the account type:

  (a) For short-term loans (one year or less), from the **Other Account Type** drop list, choose **Other current liability** as the account type.

  (b) For long-term loans, choose **Long Term Liability**.

- ▌ Click **Continue**.

- ▌ Enter the name of the lender and a description of the loan.

- ▌ Leave the opening balance at 0.00.

- ▌ Click **Save & Close**.

2. Next, set up an asset account for the truck.

- ▌ Click **Home**, then click the **Chart of Accounts** icon, or from the **Lists** menu, choose **Chart of Accounts**.

- ▌ Pull down the **Account** menu at the bottom of the window and choose **New**.

- ▌ Select the account type. Click **Continue**.

- ▌ Create the new account similar to the example in Figure 3-6.

- ▌ Click **Save & Close**.

Figure 3-6

Use the Add New Account window to create a new asset account for a truck.

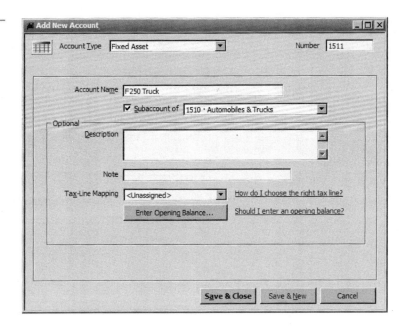

3. Now we'll set up a Fixed Asset Item for the truck. Make sure you have a vendor set up for making loan payments. For example, if you'll be paying Ford Motor Credit, get that vendor entered now. For more information on entering vendors, see Chapter 8.

   ▌ From the **Lists** menu, choose **Fixed Asset Item List**.

   ▌ Pull down the **Item** menu at the bottom of the window and choose **New**.

   ▌ Create the new account similar to the example in Figure 3-7.

   ▌ Click **OK**.

4. Next, we'll enter a transaction that records the purchase of the asset and the creation of a loan in the liability account you set up. For this example, let's say you purchased a truck for $30,000, put down $3,000, and took out a loan for $27,000.

   ▌ Click **Home**, then click the **Write Checks** icon, or from the **Banking** menu, choose **Write Checks**.

   ▌ Fill out the top portion of the check as shown in Figure 3-8.

   ▌ Click the **Expenses** tab.

   ▌ From the drop-down list in the **Account** column, select the liability account you created for the truck. In our example, it's 2511 Ford F250 Loan. In the Amount column, enter the amount of the loan as a negative number, as shown in Figure 3-8.

Figure 3-7

Use the New Item window to create a fixed asset item for a truck.

Figure 3-8

Enter the loan amount as a negative number on the Expenses tab.

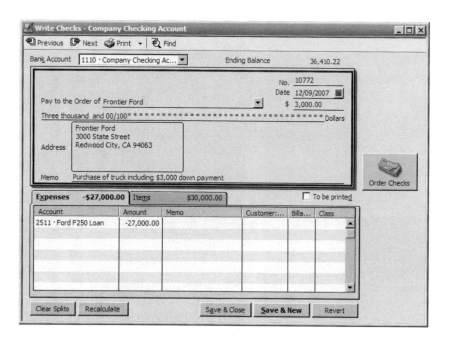

Figure 3-9

Use the Items tab of the Write Checks window to record the cost of the truck.

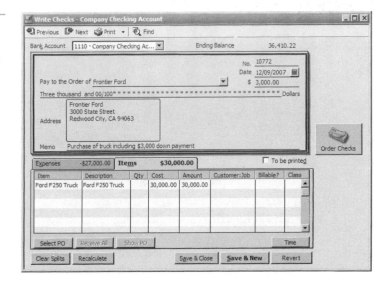

Figure 3-10

Enter the account information for the loan.

▌ Click the **Items** tab.

▌ From the drop-down list in the Item column, select the fixed asset item you created for the truck. In the Amount column enter the cost of the truck. See Figure 3-9.

▌ Click **Save & Close**.

5. From the **Banking** menu, select **Loan Manager**.

6. Click **Add Loan**.

7. Fill in the Enter account information for this loan section, as shown in Figure 3-10. Click **Next**.

Figure 3-11

Enter the payment information for the loan.

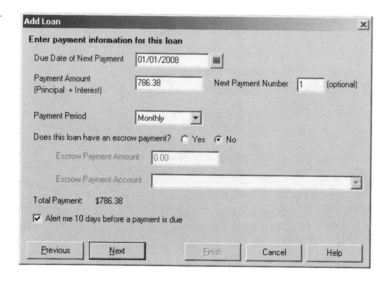

Figure 3-12

Enter the interest information for the loan.

8. Fill in the Enter payment information for this loan section, as in Figure 3-11. Click **Next**.

9. Fill in the Enter interest information for this loan section as shown in Figure 3-12.

10. Click **Finish**. The loan will now appear in the Loan Manager window. See Figure 3-13.

Making Loan Payments from Loan Manager

If you're using Loan Manager to handle your loans, you should pay all of your business loans from the Loan Manager.

Figure 3-13
Track loan-related information in the Loan Manager.

When you're ready to make a payment, click **Set Up Payment** in the Loan Manager. From that window, the Loan Manager takes you directly to the Write Checks or Enter Bills windows, where you can edit your payments.

Keep in mind that you won't be allowed to automatically create a loan payment if you haven't logged in the balance of the loan as a liability.

Cash Flow Projector

This report helps you forecast how much cash you'll have by projecting your cash inflows, cash disbursements, and bank account balances on a week-by-week basis.

To see a list of the transactions that make up an amount, double-click the amount.

You can change the report date range, you can change the forecasting periods, and you can delay receipts for late customer payments. QuickBooks Help file will step you through these tasks.

▌ To create this report, click the **Report Center**, or from the **Reports** menu, choose **Company & Financial** and then **Cash Flow Forecast**.

Forecasting

Forecasting allows you to make predictions about future revenue and cash flow, as well as assess "what if" scenarios to help you make better business decisions.

A forecast can be created from scratch, from actual data from the previous fiscal year, or from the previous fiscal year's forecast. A forecast is uniquely identified by its fiscal year, and if desired, further identified by Customer:Job or Class.

▌ To use the forecasting feature, from the **Company** menu select **Planning & Budgeting**, then **Set Up Budgets**.

Business Planning

Using your QuickBooks data and answers to simple step-by-step questions, a balance sheet, profit and loss statement, and statement of cash flows will automatically be projected for the next three years. The business plan is based on the format recommended by the U.S. Small Business Administration for loan applications or a bank line of credit.

▌ To use the Business Planning feature, from the **Company** menu, select **Planning and Budgeting**, and then **Use Business Plan Tool**.

In this chapter we explained your Chart of Accounts. In the next chapter, Items, we'll show you how to set up items that match your job phases or divisions. Then we'll show you how to link each item to an account.

Chapter 4

Items

Although QuickBooks Pro uses the name *Items*, you can think of them as job phases — like the 16 CSI divisions, HomeTech's 28 Remodeling phases, the 31 phases in the *sample.qbw* company file, or job phases you've defined for your own business. They're like containers where you store information.

Using QuickBooks Pro Items, you can break any job into measurable, trackable units. For example, you can track your payroll and material expenses for wall framing to find out if you're over your estimated costs. It's important to do that *before* it's too late to correct any problems.

Probably no two companies will use the same set of items. The *sample.qbw* file you received with this book uses items for estimating, invoicing, and job costing. Although we've put an Items List on *company.qbw*, you really can't expect our set of items to match yours. But you can easily add, delete, and edit the items to suit your job costing needs.

One important thing to understand about items is that they link to the Chart of Accounts. When you set up a job cost item, you should link it to both a job related cost account, like 5110 Job Related Costs, and an income account, like 4110 Construction Income.

To compile a project report (Job Profitability, Job Estimates, and Job Progress, for example), you need to use items. In the *sample.qbw* data, we've created several items. You can study this list to see how we use the items in estimates and invoices shown in this book. Then you can set up items to best suit your own business. To see the Items List in *sample.qbw*:

▌ Open *sample.qbw*.

▌ Click **Home**, then click **Items & Services**, or from the **Lists** menu, choose **Item List**.

The project reports built into QuickBooks Pro use items to break job cost reports into smaller units — called job phases. Job phases should be the phases you actually use to estimate a job. That way you can get reports

Name	Description	Type	Account	On Hand	Price
◆01 Plans & Permits	Plans & Permits	Service	4110 · Construction Income		0.00
◆01.1 Plans	Plans	Service	4110 · Construction Income		0.00
◆01.2 Building Permits	Building Permits	Service	4110 · Construction Income		0.00
◆01.3 City & Co. Lic's & Fees	City & County Licenses & Fees	Service	4110 · Construction Income		0.00
◆02 Site Work	Site Work	Service	4110 · Construction Income		0.00
◆02. 10 Demo	Demo	Service	4110 · Construction Income		0.00
◆03 Excavation	Excavation	Service	4110 · Construction Income		0.00
◆04 Concrete	Concrete	Service	4110 · Construction Income		0.00
◆05 Masonry	Masonry	Service	4110 · Construction Income		0.00
◆06 Floor Framing	Floor Framing	Service	4110 · Construction Income		0.00
◆07 Wall Framing	Wall Framing	Service	4110 · Construction Income		0.00
◆08 Roof Framing	Roof Framing	Service	4110 · Construction Income		0.00
◆09 Roof Flashing	Roofing Flashing	Service	4110 · Construction Income		0.00
◆10 Exterior Trim & Decks	Exterior Trim & Decks	Service	4110 · Construction Income		0.00
◆11 Siding	Siding	Service	4110 · Construction Income		0.00
◆12 Doors & Trim	Doors & Trim	Service	4110 · Construction Income		0.00
◆13 Windows & Trim	Windows & Trim	Service	4110 · Construction Income		0.00
◆14 Plumbing	Plumbing	Service	4110 · Construction Income		0.00
◆15 HVAC	Heating & Cooling	Service	4110 · Construction Income		0.00
◆16 Electrical & Lighting	Electrical & Lighting	Service	4110 · Construction Income		0.00
◆17 Insulation	Insulation	Service	4110 · Construction Income		0.00
◆18 Interior Walls	Interior Walls	Service	4110 · Construction Income		0.00
◆19 Ceilings & Cover	Ceilings & Coverings	Service	4110 · Construction Income		0.00
◆20 Millwork & Trim	Millwork & Trim	Service	4110 · Construction Income		0.00
◆21 Cabinets & Vanities	Cabinets & Vanities	Service	4110 · Construction Income		0.00
◆22 Specialty	Specialties	Service	4110 · Construction Income		0.00
◆23 Floor Coverings	Floor Coverings	Service	4110 · Construction Income		0.00
◆24 Paint	Painting	Service	4110 · Construction Income		0.00
◆25 Cleanup	Cleanup & Restoration	Service	4110 · Construction Income		0.00
◆26 Landscaping & Paving	Landscape & Paving	Service	4110 · Construction Income		0.00
◆27 Contingency	Contingency	Service	4110 · Construction Income		0.00

Item ▾   Activities ▾   Reports ▾   Excel ▾   ☐ Include inactive

Figure 4-1
The Item List window shows information about each item created in QuickBooks Pro.

comparing your estimated costs to your actual job costs. You'll find that these reports are invaluable for running a profitable business.

Entering Items for a Non-Inventory Based Business

If you don't track inventory, you can use service inventory items to divide jobs into phases of construction. For example, if you use a specific list such as the 16 CSI divisions to make estimates, you can create that list in QuickBooks Pro using service items.

Let's see how the company on *sample.qbw*, Twice Right Construction, did it. Twice Right computerized their estimating and job costing using QuickBooks Pro items. Figure 4-1 shows some of their QuickBooks Pro items.

Figure 4-2

Use the New Item window to create a new service item.

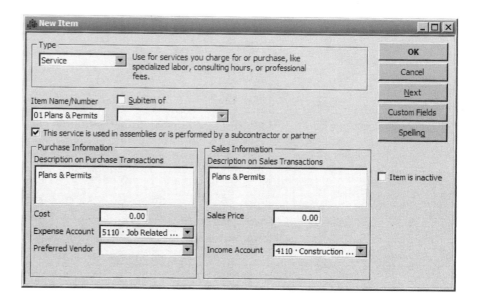

To create a new service item:

▮ Click **Home**, then click **Items & Services**, or from the **Lists** menu, choose **Item List**. In the Item List window, pull down the **Item** menu and choose **New**. See our example in Figure 4-2.

▮ For **Type**, select **Service**. A "service" type will allow you to post labor, materials, subcontractor costs, etc. to it.

▮ Make sure you select the box in front of **This service is used in assemblies or is performed by a subcontractor or partner**. *This is important.* You need to select this box and the exact same accounts each time you use the item to make sure your accounting is accurate. Notice that the cost of the item is linked to the expense account you're using for Job Related Costs (5110 in our example). The price you're going to charge the customer will be linked to your Construction Income account (4110 in our example).

▮ Click **OK** to create the new item.

You can use subitems to do more in-depth job cost analysis. To set up a subitem, you have to create or select an item and then create its subitems. For example, using the CSI divisions to track Demo and Excavation costs as items, you could set up an item 02 Site Work. Then you can set up a subitem called 02.10 Demo.

To create a subitem:

▮ Click **Home**, then click **Items & Services**, or from the **Lists** menu, choose **Item List**. In the Item List window, pull down the **Item** menu and choose **New**.

Figure 4-3

You can create subitems to get a more detailed breakdown of an item.

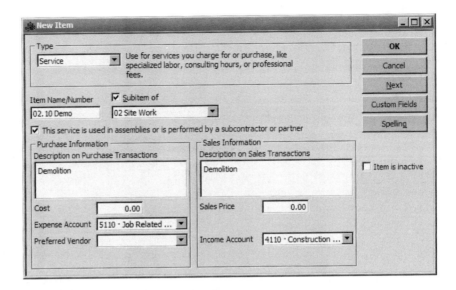

- For **Type**, select **Service**. A "service" type will allow you to post labor, materials, subcontractor costs, etc. to it.

- Select **This service is used in assemblies or is performed by a subcontractor or partner**. This will allow us to track estimated costs and actual costs for this item.

- Click **Subitem of** then select the new item this will be a subitem of. In our example, shown in Figure 4-3, we create the subitem 02.10 Demo of the item 02 Site Work.

- Fill in all other fields the same as you would for any item.

- Click **OK** to create the subitem.

Entering Items for an Inventory Based Business

Most general contractors and remodeling contractors don't have, or track, inventory (and in fact we discourage them from doing so). But if you're an electrical, mechanical, roofing subcontractor, or a specialty contractor you'll probably have a use for tracking inventory.

You can set up inventory items to track the materials you stock so you'll know when it's time to reorder them. You can keep track of how many parts are in stock after a sale, how many you've ordered, the cost of the goods you've sold, and the value of your inventory. You'll use purchase orders to track the materials you've bought, and invoices to record sales of materials. Just remember to use the same inventory part item on both invoices and purchase orders.

Figure 4-4

Use the New Item window to create a new inventory item.

To create a new inventory item:

▌ Click **Home**, then click **Items & Services**, or from the **Lists** menu, choose **Item List**. In the Item List window, pull down the **Item** menu and choose **New**.

▌ For **Type**, select **Inventory Part**. See our example in Figure 4-4. If Inventory Part doesn't show up in your pull-down list under **Type**, then you'll need to change your Purchases & Vendors Preferences. To do that, from the **Edit** menu, choose **Preferences**, click **Items & Inventory**, click the **Company Preferences** tab, and select **Inventory and purchase orders are active**. Click **OK**.

▌ In the New Item window, in **Item Name/Number**, enter a short reference to uniquely identify the inventory item.

▌ In **Description on Purchase Transactions**, enter any description you like (but remember the description is automatically placed on a purchase order). For example, some people use the vendor's part number and description here to help the vendor identify what you're buying.

▌ In **Cost**, enter the cost of the item.

▌ In **COGS Account**, enter the appropriate Cost of Goods Sold account.

▌ In **Preferred Vendor**, enter the vendor you usually order the material from.

▌ In **Description on Sales Transactions** enter a description. This description prefills on your invoices automatically.

■ In **Sales Price**, enter the amount you charge a customer for the item.

■ In **Income Account**, select the appropriate account. You may need to create a new account.

■ In **Inventory Information**, enter the amounts for **Reorder Point** and **On Hand**. If this is a new item in your inventory, leave **On Hand** and **Total Value** at zero. If this item is already in your inventory, you'll have to fill in these fields to get the inventory up to current levels.

■ Click **OK** to create the new item.

Creating a Group of Items

You can group items together and refer to them as one unit. For example, you could make a "Plans & Permits" group item by grouping the items for plans, building permits and city license fee. Each item in the Plans & Permits group item has its own cost and selling price. When you choose an item for your estimate or invoice, you simply have to enter a quantity to get the total price for the Plans & Permits group item.

Suppose you want to use two formats for estimates: one that's an expanded detail view for your own job costing, and another that's a condensed version in a summary format that you can print out for a customer. This is especially useful if you've created a large number of subitems for each of your primary items. The condensed format might be appropriate if you're estimating the job on a fixed contract price and don't want to give the customer too much detail.

To create a new group item:

■ Click **Home**, then click **Items & Services**, or from the **Lists** menu, choose **Item List**. In the Item List window, pull down the **Item** menu and choose **New Item**. See our example in Figure 4-5.

■ In **Type**, select **Group**. In **Group Name/Number**, enter a name for the new group. In **Description**, enter a short description of the group you're creating.

■ Select **Print items in group** if you want all the items printed when you print the group item on an invoice. This will list each item in the group individually on the invoice. If you don't select this, only the group item name will appear on the invoice.

■ Click in the column titled **Item**. Click the down arrow that appears and select the first item you want to include in the group. If you don't enter a quantity for the item, a value of 1 will be assumed.

■ Click other items as you wish.

■ Click **OK** to create the group item.

Figure 4-5

Use grouped items to make estimating and invoicing quick and efficient.

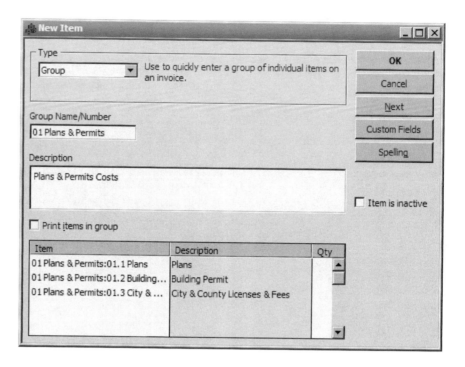

After you've created a group item, you can edit it if you need to. Any item can be in more than one group. To edit a group item:

▌ Click **Home**, then click **Items & Services**, or from the **Lists** menu, choose **Item List**.

▌ In the Item List window, double click the group item you want to edit and make changes in the Edit Item window.

▌ Click **OK**.

Entering Non-Job Related Items

In the Item List in the *sample.qbw* data file, you'll notice that near the end of the list we have several items that aren't job related. These are items that you may or may not use, such as Deposit, Beginning Balance (Beg Bal), Subtotal and Markup. You can use the Deposit item if you accept deposits from customers *and* when you subtract the deposits from the billing. For more information on customer deposits, and using subtotal items, see Chapter 13. Figure 4-6 shows how we created the Deposit item.

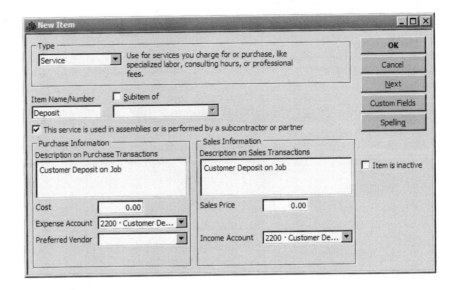

Figure 4-6

Set up the Deposit item as a template for customer deposits.

You can use the Beg Bal item when you enter your outstanding invoices — both Accounts Receivable and Accounts Payable. Figure 4-7 shows how we created the Beg Bal item.

You can use the Subtotal item if you'll be billing your customers on a time and materials basis.

- Click **Home**, then click **Items & Services**, or from **Lists** menu, select **Item List**.

- In the Item List Window, click on the **Item** pull-down menu and select **New**.

- In **Type**, select **Subtotal**, see Figure 4-8.

- In **Item Name/Number**, type Subtotal.

- Click **OK** to create the Subtotal Item.

The Markup item is used if you've been billing your customers on a time and materials basis and want to add a separate line item for Markup.

- Click **Home**, then click **Items & Services**, or from **Lists** menu, select **Item List.**

- In the Item List Window, click the **Item** pull-down menu and select **New.**

- In **Type**, select **Other Charge**, see Figure 4-9.

- In **Item Name/Number**, type Markup.

- Click **OK** to create the Markup Item.

After you finish setting up the items you need, you're ready to go on to Chapter 5 to set up payroll items.

Figure 4-7

Use the Beginning Balance item to enter outstanding invoices for Accounts Receivable and Accounts Payable.

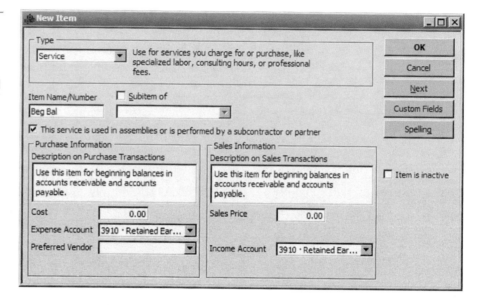

Figure 4-8

The Subtotal item is used on a time and materials invoice in Chapter 13.

Figure 4-9

Create a Markup item if you will be creating a time and materials invoice, as in Chapter 13.

Chapter 5

Payroll Items

In QuickBooks Pro, you must set up payroll items to track wages, employer and employee payroll taxes, employee and employer paid benefits, and any deductions from an employee's check. In this chapter, we'll show you how to set up the payroll items you need. In Chapter 15 (Payroll), we'll show you how to actually use the payroll items to process your payroll.

First, we'll show you how to set up payroll items to track gross wages by workers' compensation classification. This will help you prepare your workers' comp report. Second, we'll show you how to set up employer contributions to track how much you owe the insurance company for workers' comp. We'll also track the cost of the insurance by job. Third, we'll show you how to change the existing payroll items if you *don't* want to track workers' comp in QuickBooks Pro. Finally, we'll set up an hourly wage payroll item for tracking your costs to a job if you're a sole proprietor or partner of your company.

Throughout the chapter, we'll show you how to link each payroll item to the correct account in your Chart of Accounts to make sure your financial reports will be accurate. The accounts used in our examples are from the Chart of Accounts in the data file *sample.qbw* on the CD in the back of this book. If your Chart of Accounts is different, be sure to link the payroll item to the correct account.

Using Payroll Items to Track Workers' Comp Costs

Workers' comp insurance covers work-related accidents, and it's mandatory in some states. Usually, the insurance company will require you to track gross wages paid according to their classification system. They'll give you the name and rate for each classification. This insurance can be very costly for your business. For example, the roofing classification can be as high as 54 percent of wages. This means that for every $100 you pay an employee working on roofing, you'll have to pay your insurance carrier $54 for workers' comp insurance coverage. It's critical that you show these costs in your job cost reports.

To track workers' comp costs, you need to set up a payroll item for each workers' comp classification your company uses. For example, if your workers'

Figure 5-1

This Payroll Item List displays information about the payroll items created in QuickBooks Pro.

Item Name	Type ▲	Amount	Annual L...	Tax Tracking	Payable To	Account ID
Officer's Salary	Yearly Salary			Compensation		
1027 - Design	Hourly Wage			Compensation		
5028 - Masonry	Hourly Wage			Compensation		
5188 - Plumbing	Hourly Wage			Compensation		
5190 - Electrical	Hourly Wage			Compensation		
5213 - Concrete	Hourly Wage			Compensation		
5475 - Painting	Hourly Wage			Compensation		
5553 - Roofing	Hourly Wage			Compensation		
5646 - Carpentry	Hourly Wage			Compensation		
9891 - Office Labor	Hourly Wage			Compensation		
9891 - Sick Hourly Rate/Office	Hourly Wage			Compensation		
9891 - Vac Hourly Rate/Office	Hourly Wage			Compensation		
9899 - Supervision	Hourly Wage			Compensation		
Officer's Direct Labor	Hourly Wage			Compensation		
Owner's Labor	Hourly Wage			Compensation		
Bonus - Indirect	Addition			Compensation		
Bonus - Officer's	Addition			Compensation		
Employee Advance	Deduction			None		
Medical deduction	Deduction	-50.00		None		
Owner's Labor Deduction	Deduction			Compensation		
1027-Design Costs	Company Contribution	0.9%		None	State Compensation I...	
5028-Masonry Costs	Company Contribution	7.3%		None	State Compensation I...	
5188-Plumbing Costs	Company Contribution	9.1%		None	State Compensation I...	
5190-Electrical Costs	Company Contribution	10.2%		None	State Compensation I...	
5213-Concrete	Company Contribution	10.0%		None	State Compensation I...	
5475-Painting Costs	Company Contribution	4.2%		None	State Compensation I...	
5553-Roofing Costs	Company Contribution	24.2%		None	State Compensation I...	
5646-Carpentry Costs	Company Contribution	10.2%		None	State Compensation I...	
9891-Office Labor Costs	Company Contribution	0.9%		None	State Compensation I...	
9899-Supervision Costs	Company Contribution	10.12%		None	State Compensation I...	9807-0978
Advance Earned Income Credit	Federal Tax			Advance EIC P...	Wells Fargo Bank	94-4555...
Federal Unemployment	Federal Tax	0.8%	7,000.00	FUTA	Internal Revenue Ser...	94-4555...
Federal Withholding	Federal Tax			Federal	Wells Fargo Bank	94-4555...
Medicare Company	Federal Tax	1.45%		Comp. Medicare	Wells Fargo Bank	94-4555...
Medicare Employee	Federal Tax	1.45%		Medicare	Wells Fargo Bank	94-4555...
Social Security Company	Federal Tax	6.2%	97,500.00	Comp. SS Tax	Wells Fargo Bank	94-4555...

Payroll Item ▼   Activities ▼   Reports ▼   ☐ Include inactive

comp provider uses a classification of 5028, you can set up an hourly rate payroll item with the same code number and call it 5028 Masonry. Then, when you enter hours for an employee who did masonry work, you choose that payroll item.

When you run a payroll report, you'll get the payroll broken down by each workers' comp classification you set up with a payroll item.

Setting Up a Payroll Item for a Workers' Comp Classification

Here's how to set up an hourly wage payroll item for a workers' comp wage classification:

▎ From the **Lists** menu, choose **Payroll Item List**. See Figure 5-1.

▎ A window may appear asking if you would like to learn more about QuickBooks Payroll Services. Select **No**.

▎ Click **Payroll Item** and choose **New**.

▎ In the Select Setup Method window, choose **Custom Setup**.

▎ Click **Next**.

▎ In Payroll item type window, select **Wage (Hourly Wages, Annual Salary, Commission, Bonus)**, as shown in Figure 5-2.

▎ Click **Next**.

Figure 5-2

You need to create an hourly wage payroll item for each workers' comp classification.

Figure 5-3

In this example, we've specified hourly wage for the workers' comp payroll item we're entering.

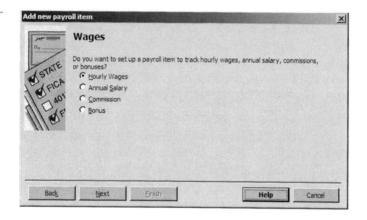

Figure 5-4

Select regular pay for each workers' comp payroll item.

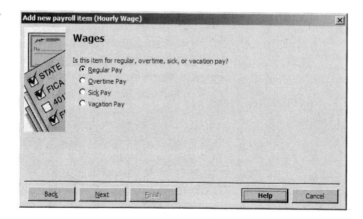

▌ In the Wages window, select **Hourly Wages**, as shown in Figure 5-3.

▌ Click **Next**.

▌ In the next Wages window, select **Regular Pay**, as shown in Figure 5-4.

▌ Click **Next**.

Figure 5-5

It is important to accurately classify each payroll item by name.

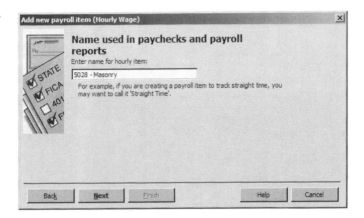

Figure 5-6

In this example, 5210 - Job Labor (Gross Wages) is the correct expense account.

▌ In the Name used in paychecks and payroll reports window, enter the name of the workers' comp classification. In our example, it's 5028 - Masonry. See Figure 5-5.

▌ Click **Next**.

▌ In the Expense account window, enter the account number for tracking this expense. In our example, it's 5210 - Job Labor (Gross Wages). See Figure 5-6.

▌ Click **Finish**.

After you enter the new payroll item, it will be added to your Payroll Item List.

Setting Up a Payroll Item for a Company Contribution to a Workers' Comp Classification

Along with the payroll item 5028 - Masonry, you also need to set up a company contribution payroll item for 5028 - Masonry Costs. Here's how to do that:

Figure 5-7

You also need to set up a Company Contribution payroll item type when you set up a workers' comp classification payroll item.

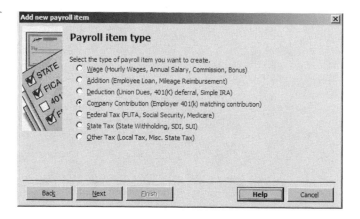

Figure 5-8

Use this window to enter the name of the company contribution payroll item type for the workers' comp classification.

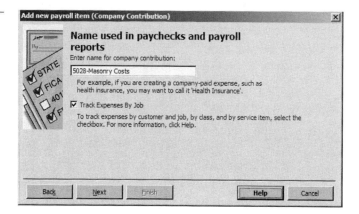

▌ From the **Lists** menu, choose **Payroll Item List**.

▌ In the Payroll Item List window, pull down the **Payroll Item** menu and choose **New**.

▌ In the Select Setup Method window, choose **Custom Setup**.

▌ Click **Next**.

▌ In the Payroll item type window, select **Company Contribution (Employer 401(k) matching contribution)**. See Figure 5-7.

▌ Click **Next**.

▌ In the Name used in paychecks and payroll reports window, enter your workers' comp code and code name. We entered 5028 - Masonry Costs.

Tip! The name of the company contribution has to be slightly different from the name of the wage payroll item you entered previously. See Figure 5-8.

▌ Be sure to select **Track Expenses By Job** to get workers' comp costs in your job costs reports. Each employee's workers' comp cost will be divided by the jobs the employee worked on during the pay period.

▌ Click **Next**.

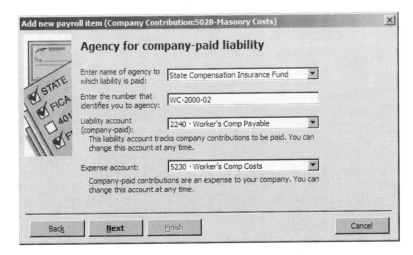

Figure 5-9

This window shows the details of the payee, the account number, and the correct liability and expense accounts for proper accounting.

▌ In the Agency for company-paid liability window, in the first box, enter your workers' comp insurance carrier's name. In our example, it's State Compensation Insurance Fund. See Figure 5-9.

▌ In the second box, enter the number that identifies you to the agency (e.g. an insurance policy number).

▌ In the third box, enter your workers' comp liability account. In our example, it's 2240 - Workers' Comp Payable.

▌ In the fourth box, enter your expense account. In our example, it's 5230 - Workers' Compensation Costs.

▌ Click **Next**.

▌ In the Tax tracking type window, since this item is not reported on your tax forms, select **None** from the pull-down list. See Figure 5-10.

▌ Click **Next**.

▌ In the Taxes window, don't select any taxes. Workers' comp company-paid cost isn't subject to any additional federal or state taxes. See Figure 5-11.

▌ Click **Next**.

▌ In the Calculate based on quantity window, Select **Calculate this item based on quantity** to multiply each employee's gross wages by the percentage rate your insurance company charges you for the workers' comp classification. See Figure 5-12.

▌ Click **Next**.

▌ In the Default rate and limit window, enter the rate for the workers' comp code in the first box. For example, we entered 9.1%. This is the rate we pay when an employee works on masonry tasks. See Figure 5-13. The rate is given to you by your workers' comp insurance

Figure 5-10

Choose the option "None" from the tax classification list for an item you don't report on your tax forms.

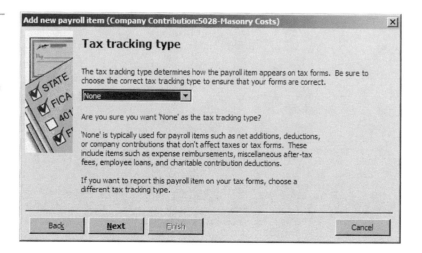

Figure 5-11

Leave these items blank because workers' comp company-paid costs aren't taxed.

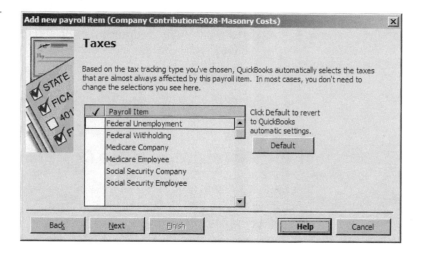

Figure 5-12

Select Based on Quantity to multiply employee gross wages by your insurance company's percentage rate.

Figure 5-13

Enter the correct percentage rate for the workers' comp classification.

company. The rates for each company contribution will need to be updated when your insurance carrier updates (increases or decreases) your workers' comp rates.

▌ Click **Finish** to complete the new payroll item.

▌ Click **OK** to set up a scheduled payment for this payroll item.

Using Payroll Items If You Don't Track Workers' Comp

If you don't want to track workers' comp, you can add new payroll items or change the existing default payroll items to your own classifications. We suggest you add new payroll items because the default items don't default to the correct account in the Chart of Accounts. We want to make sure field labor is correctly allocated to job costs and office labor is correctly allocated to an overhead expense. We suggest you use these payroll items: Field Labor, Field Labor OT, and Office Labor.

Adding a New Payroll Item

You should enter a new wages payroll item for overhead or office payroll because you'll want to capture the costs associated with your office payroll. To add a new item for Office Labor:

▌ From the **Lists** menu, choose **Payroll Item List**.

▌ In the Payroll Item List window, pull down the **Payroll Item** menu and choose **New**.

▌ In the Select Setup Method window, choose **Custom Setup**.

▌ Click **Next**.

▌ In the Payroll item type window, select **Wage** (**Hourly Wages, Annual Salary, Commission, Bonus**), as shown in Figure 5-14.

Figure 5-14

Select a Wage item type for the Office Labor payroll item type.

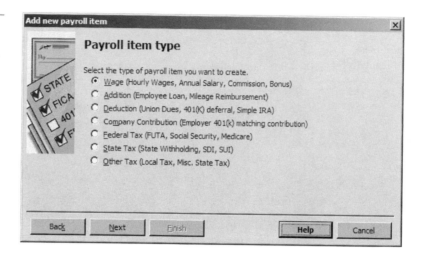

Figure 5-15

Select Hourly Wages for the Office Labor payroll item.

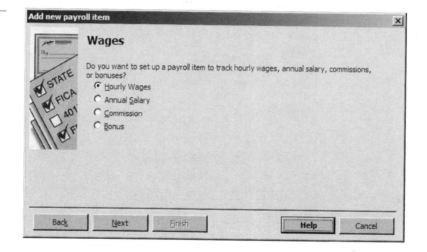

▌ Click **Next**.

▌ In the Wages window, select **Hourly Wages**, as shown in Figure 5-15.

▌ Click **Next**.

▌ In the next Wages window, select **Regular Pay**, as shown in Figure 5-16.

▌ Click **Next**.

▌ In the Name used in paychecks and payroll reports window, enter the name of the overhead labor item. In our example shown in Figure 5-17, it's Office Labor.

▌ Click **Next**. If you type a name that is already a payroll item, a Warning box will appear requesting you use a different name.

Figure 5-16

Select Regular Pay for the Office Labor payroll item.

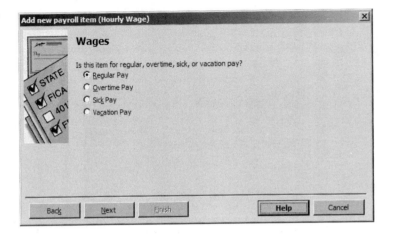

Figure 5-17

Enter the name of the Office Labor payroll item here.

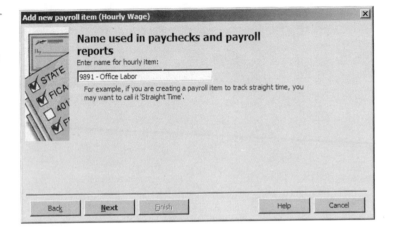

▌ In the Expense account window, enter the account number for tracking this expense. It's 6501 - Payroll (office staff) in our example, shown in Figure 5-18. Be sure you select the proper expense account from your Chart of Accounts here. The Expense Account you select here links the payroll item to your Chart of Accounts.

▌ Click **Finish**.

Including Sole Proprietor and Partners' Time Costs in Job Costs

If your company is organized as a sole proprietorship or a partnership, you'll be taking draws instead of issuing payroll checks for yourself or your partner. Because of this, QuickBooks Pro reports won't show any costs for your or your partner's time.

Figure 5-18

In this example, 6501 - Payroll (office staff) is the correct expense account for the Office Labor payroll item.

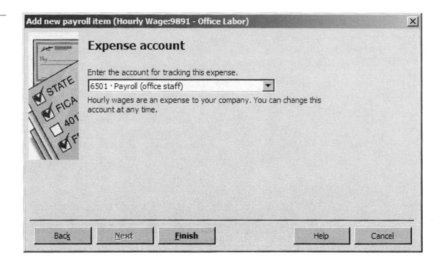

To get these costs into any job cost report, you have to transfer owner or partner time (in dollars) using an hourly wage payroll item and a deduction payroll item. In Chapter 15, Payroll, we'll show you how to actually use the hourly wage item to enter owner or partner time on a timesheet, process payroll, and then deduct the payroll back out of your accounting using the deduction payroll item.

In this section, we'll just show you how to set up the two payroll items you need — Owner's Labor and Owner's Labor Deduction.

Setting Up an Hourly Wage Payroll Item for an Owner or Partner

▌ From the **Lists** menu, choose **Payroll Item List**.

▌ In the Payroll Item List window, pull down the **Payroll Item** menu and choose **New**.

▌ In the Select Setup Method window, choose **Custom Setup**.

▌ Click **Next**.

▌ In the Payroll item type window, select **Wage (Hourly Wages, Annual Salary, Commission, Bonus)** as shown in Figure 5-19.

▌ Click **Next**.

▌ In the Wages window, select **Hourly Wages** as shown in Figure 5-20.

▌ Click **Next**.

▌ In the next Wages window, select **Regular Pay**, as shown in Figure 5-21.

▌ Click **Next**.

Figure 5-19

Select a Wage item type for the owner/partner payroll item.

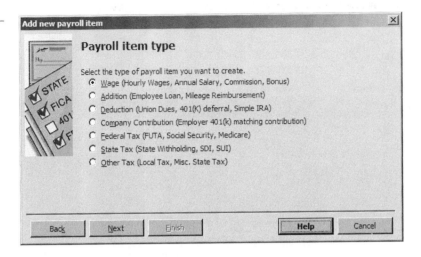

Figure 5-20

Select Hourly Wages for the owner/partner payroll item.

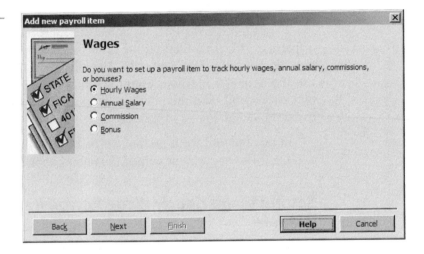

Figure 5-21

Select Regular Pay for the owner/partner payroll item.

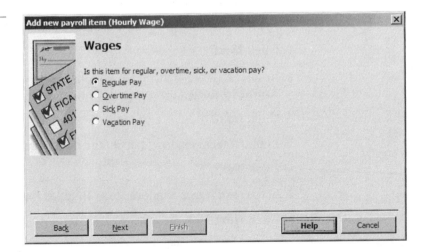

Figure 5-22

Enter the name of the owner/partner payroll item here.

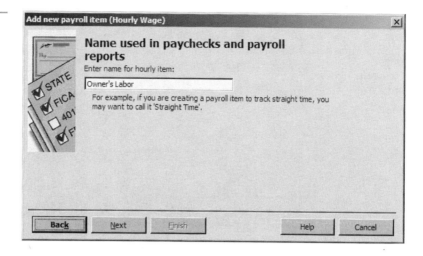

Figure 5-23

It's important to enter the appropriate expense account number to maintain correct accounting records. In this example, 3999 - Owner's Time to Jobs is the correct account.

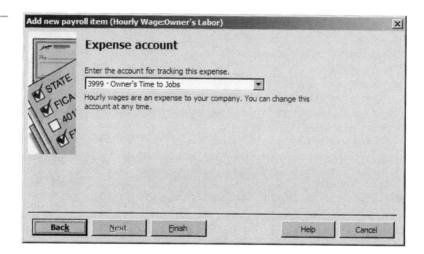

In the Name used in paychecks and payroll reports window, enter the name of the overhead labor item. In our example shown in Figure 5-22, it's Owner's Labor.

Click **Next**.

In the Expense account window, enter the account number for tracking this expense. It's 3999 - Owner's Time to Jobs, in the example in Figure 5-23. Be sure you select the proper expense account from your Chart of Accounts here. The Expense Account you select here links the payroll item to your Chart of Accounts.

Click **Finish**.

Figure 5-24

You also need to set up a Deduction payroll item type when you set up an owner/partner payroll item.

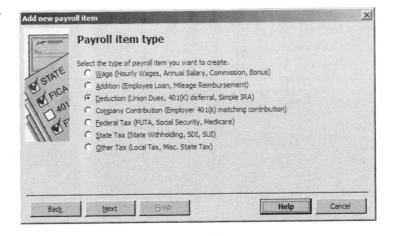

Figure 5-25

Enter the Owner's Labor Deduction in the name field.

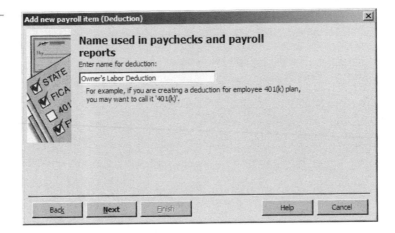

Setting Up a Deduction Payroll Item for an Owner or Partner

When you set up this deduction item, you'll be able to reverse wages and payroll taxes from the owner's check.

▌ From the **Lists** menu, choose **Payroll Item List**.

▌ In the Payroll Item List window, pull down the **Payroll Item** menu and choose **New**.

▌ In the Select Setup Method window, choose **Custom Setup.**

▌ Click **Next**.

▌ In the Payroll item type window, select **Deduction (Union Dues, 401(K) deferral, Simple IRA)**, as shown in Figure 5-24.

▌ Click **Next**.

Figure 5-26

Leave the Name field blank, enter a short description in the Number field, and in the Liability Account field enter account number 3999 - Owner's Time to Jobs.

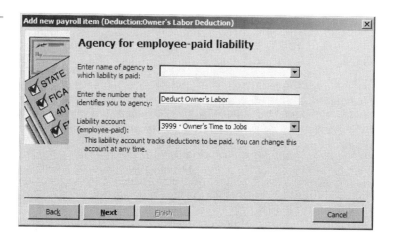

Figure 5-27

Select Compensation from the tax classification list for the Owner's Labor deduction payroll item.

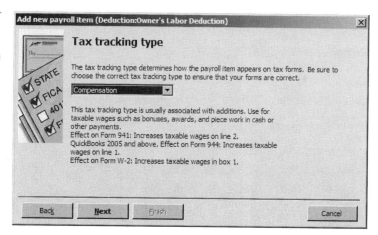

▌ In the Name used in paychecks and payroll reports window, enter the name of the deduction item. In our example shown in Figure 5-25, it's Owner's Labor Deduction.

▌ Click **Next**.

▌ In the Agency for employee-paid liability window, leave the first box blank. In the second box, in this case, you wouldn't use a number, just a description of how you're using this payroll item. In our example, we've entered Deduct Owner's Labor, as shown in Figure 5-26.

▌ In Liability account, enter 3999 - Owner's Time to Jobs.

▌ Click **Next**.

▌ In the Tax tracking type window, select **Compensation** from the pull-down list, as shown in Figure 5-27. This is important!

Figure 5-28

Leave the tax boxes blank because there's no tax associated with the Owner's Labor deduction.

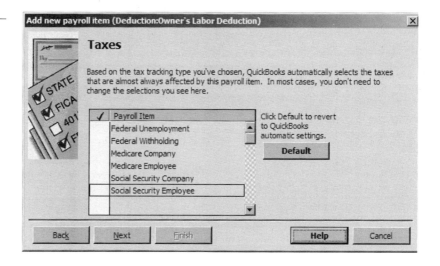

Figure 5-29

Select Neither when you set up the deduction payroll item for owner/partner labor.

▮ Click **Next**.

▮ In the Taxes window, because this deduction isn't associated with any taxes, don't select anything, just click **Next**. QuickBooks Pro will give you the Payroll Item Taxability warning box. Ignore the warning and click **Yes**. See Figure 5-28.

▮ In the Calculate based on quantity window, select **Neither**. See Figure 5-29.

▮ Click **Next**.

Figure 5-30

Select gross pay for the Owner's Labor Deduction payroll item.

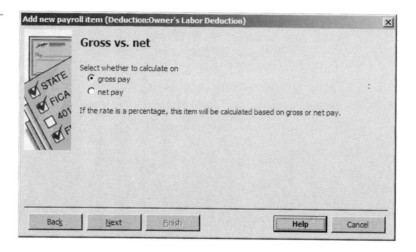

Figure 5-31

Use a default rate of 100% to automatically deduct the owner's time.

▌ In the Gross vs. net window, select **gross pay**. See Figure 5-30.

▌ Click **Next**.

▌ In the Default rate and limit window, enter 100% to automatically deduct 100% of the owner's time. See Figure 5-31.

▌ Uncheck the annual limit box.

▌ Click **Finish**.

Keep in mind that this chapter just showed you how to set up payroll items. Later in the book, in Chapter 15, we'll show you how to use the payroll items you set up here to process a payroll.

After you set up your payroll items, you're ready to move on to setting up classes.

Classes are one of the "containers" QuickBooks Pro gives you to earmark transactions and retrieve them later in useful reports. Classes, in fact, are accessible from almost every transaction in QuickBooks Pro. So, when you fill in timecards, you can pick a class to "aim" the time to. Or if you're filling in an invoice, you can pick a class to toss the line item cost into.

Using Classes to Track Cost Categories

We recommend that you use classes to track Labor, Materials, Subcontractors, Equipment Rentals, and Other.

Your tax preparer can use class reports to get information to prepare your tax return. It doesn't matter whether you're a sole proprietor, partnership, or corporation, your tax return has a section for Cost of Goods Sold. In a construction company, that's a job related cost. The Cost of Goods Sold section on the tax return asks for a summary of your job costs by Labor, Materials, Subcontractors, Equipment Rental, and Other. If you break out your classes into these categories, you can easily produce the reports your tax preparer will need.

We've included a memorized report named "Job Cost Class Report" in the *sample.qbw* file. This report shows total Revenue, Labor, Materials, Subcontractors, and Other subtotaled for the year. Figure 6-1 shows part of a report on the class "Material Costs - Job Related." In Chapter 17 (Reports), we'll show you how to access this report.

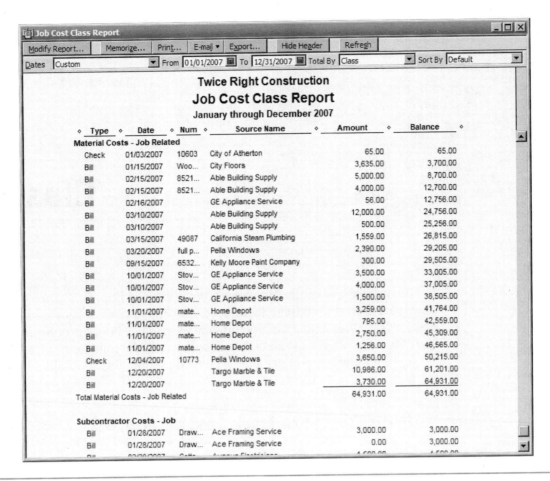

Figure 6-1
Part of a memorized Job Cost Class Report on the class Material Costs - Job Related.

How to Create a Class

Before we tell you how to create a class, you need to make sure the Class Tracking Preference is turned on. To do this:

▌ From the **Edit** menu, choose **Preferences** and click the **Accounting** icon.

▌ On the Company Preferences tab, select **Use class tracking**.

▌ Click **OK**.

Now to create a class:

▌ From the **Lists** menu, choose **Class List**. See Figure 6-2.

▌ In the Class List window, pull down the **Class** menu and choose **New**.

▌ Enter a name for the new class.

Figure 6-2

You can use the Class List provided in the company.qbw file as a model for your business.

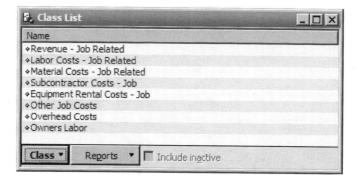

Figure 6-3

You can create subclasses to get a more detailed breakdown of a class.

If you want more detailed information — a breakdown on labor, perhaps — you can define this new class as a subclass of another class. For instance, you might want to track a specific type of labor, such as trim carpentry, to see if it's coming in as estimated. Just tell your trim carpenters to accurately record the time they spend doing casing and base, and enter the time in a subclass you set up to track that task. Then you can pull up a report that calculates the total amount of time/money spent doing casing and base. If you divide the total amount of casing and base installed by the cost of doing it, you'll get a unit cost for placing casing and base. The next time you bid casing and base, you'll have a better unit cost and should be able to make more money on that phase of your business.

To create a subclass as shown in Figure 6-3:

▌ Click the box in front of **Subclass of** and enter the appropriate class name.

You'll also notice Class is inactive in the New Class window. If you don't want a class to appear in the Class List, you can check this field to keep it from appearing in the list. To make the class active again, click Include inactive in the Class List window. Inactive classes are marked with a small "x" icon beside the class name. Click this icon to activate the class again.

▌ Click **OK** to create the new class as a subclass.

After you finish setting up the classes you want to use, move on to the next chapter where we'll show you how to set up customers and jobs.

Chapter 7

Customers and Jobs

QuickBooks Pro tracks jobs by customer — but it's important to understand that a customer can have more than one job. You can set up a main job, then track changes by setting up additional jobs using a change order number. Or you may do more than one job for a customer and want to keep them separate. Then you can assign the jobs different names, such as Rachel Olsen: Bathroom Addition and Rachel Olsen: Guest Cottage. First set up the customer, then add the job or jobs.

How to Set Up a Customer

QuickBooks Pro keeps your customer information in the Customer Center. See Figure 7-1.

Use the Customer Center to keep track of a customer's billing and shipping address, and business, home, and fax telephone numbers. The billing address might be the existing residence and the shipping address might be the address of the new residence if you're building a new home for the client.

To enter a customer:

▌ Click the **Customer Center** icon on the menu bar, or from the **Customers** menu, choose **Customer Center**.

▌ In the Customer Center window, click **New Customer & Job**, then **New Customer**.

▌ Enter the customer name. See Figure 7-2. Enter the last name first if you'd like to sort your list alphabetically by last name. Fill in the rest of the fields in the **Address Info** tab.

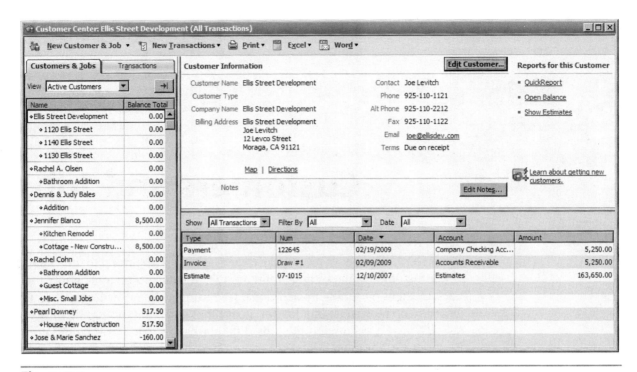

Figure 7-1

The Customer Center window displays all your existing customers and their associated jobs.

Figure 7-2

Use the Address Info tab to make additions or changes to a customer's name, contact, phone, or address.

Figure 7-3

The Additional Info tab lets you track more details about a customer. You can even create custom fields specific to your needs here.

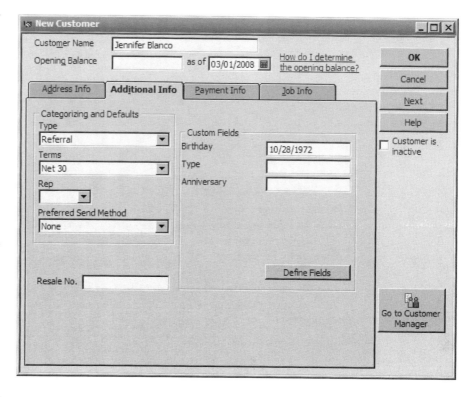

The Additional Info Tab

In the Additional Info tab window, you'll find drop-down lists for Customer Type, Terms, and Rep. You can use Customer Type as a "container" to track customers based on something that's meaningful to you. For example, you could record how the customer got your company's name — newspaper ad, referral, Chamber of Commerce, etc. See Figure 7-3.

▌ In **Type**, select the method your customer used to get your name (used to track marketing efforts). In our example, we used **Referral**.

▌ In **Terms**, select the terms of credit you wish to offer the customer and what you want printed on the invoice.

▌ In **Rep**, you can assign a salesperson to the customer and track sales by representative. If you don't have salespersons, enter the project manager's initials to track who is responsible for the job. The rep's initials can be printed on invoices.

You can also create and use custom fields in this window. You can track information about a customer, such as their birthday or anniversary, so you can do some "target" marketing with a birthday card or gift. You might also use this section to record the type of client. Some contractors track whether a client is "from Hell" or "from Heaven." Naturally, those categorized as "from Hell" are charged more for your efforts.

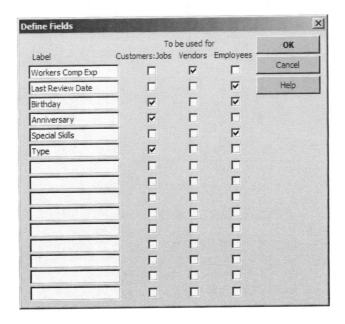

Figure 7-4

Use the Define Fields window to enter a label for a custom field and specify if it will be used for customers, vendors, or employees.

To create a custom field:

▌ In the Custom Fields area, click **Define Fields**. See Figure 7-4.

▌ In the Define Fields window, under **Label**, enter the field names you want to create.

▌ In the columns to the right, check the boxes to select where you'll be using this field (Customers:Jobs, Vendors, Employees).

▌ Click **OK** to create the new fields.

The Payment Info Tab

You can enter customer payment information in the Payment Info tab window. See Figure 7-5. If you require your customers to pay by credit card (very rare in construction), you can enter the customer's credit card information on this screen. If you enter your customer's credit card number and expiration date in QuickBooks Pro, you should immediately set up password protection and make sure that only a trusted few have access to viewing customer payment information.

▌ Click the **Payment Info** tab of the New Customer window to fill in more information about a customer.

▌ In **Account No.**, enter the account number or the PO number your customer requested you put on each invoice you send to them.

Figure 7-5

Enter customer payment information in the Payment Info tab window.

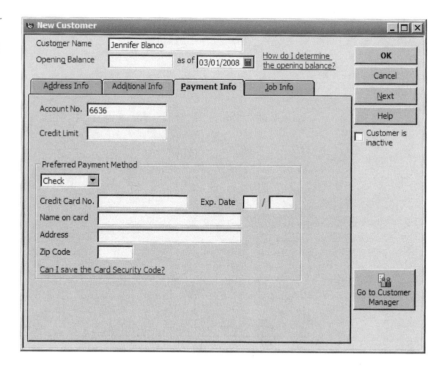

■ We suggest that you skip the Job Info tab when you're entering information for a new customer. We'll cover it in the next section.

■ Click **OK** to enter the new customer.

Adding a Job for a Customer

Whether you do one or more jobs for a customer, it's a good idea to set up each job separately. For example, if Twice Right remodels a kitchen for Jennifer Blanco, and then later builds her a guest cottage, both the kitchen and the guest cottage are separate jobs, but they should both be attached to the same customer — Jennifer Blanco. To add a job for a customer:

■ In the Customer Center window, right click the customer's name and choose **Add Job**.

■ Enter a job name in the box next to **Job Name** on the New Job window. In our example, we entered Cottage — New Construction.

■ Change information on the **Address Info** and **Additional Info** tabs if you need to.

■ Click the **Job Info** tab, shown in Figure 7-6.

Figure 7-6

Use the Job Info tab to record all the pertinent information for a particular job.

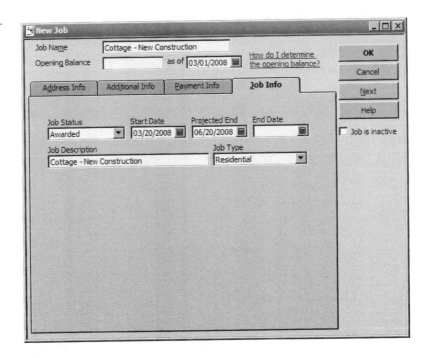

The Job Info Tab

You can enter Job Status, Start Date, Projected End, End Date, Job Description, and Job Type on the Job Info tab.

The Job Status box lets you track the status of your jobs. Pending, Awarded, In progress, Closed, Not awarded, and None are your choices. You can do a report on jobs Pending, or jobs Not awarded. A report on Not awarded vs. all your jobs will tell you the ratio of jobs you're getting to the jobs you're bidding. Use this information when you create your marketing budget. Figure 7-6 shows a Job Info tab window.

▌ From the pull-down Job Status list, choose the status of the job.

▌ In **Start Date**, enter the start date for the job.

▌ In **Projected End**, enter the projected end date for the job.

▌ In **End Date**, enter the actual date the job ended. (You'll need to come back later and fill this in.)

▌ In **Job Description**, enter the description for the job. You would fill this in if you use job numbers instead of names in the Job Name field.

The Job Type box in the Job Info tab window has a drop-down list from which you can choose a job type. For example, you could set up a job type such as Residential, Commercial, or Remodel, with subtypes for Contract or Speculation. Then, whenever you get a new job, you can choose which

Figure 7-7

You specify a name for a new job type in the New Job Type window.

Figure 7-8

Use subtypes to get a detailed breakdown of a job type.

job type the job falls under. Later (at year's end, for example) you can sort jobs by type to make a report that evaluates profitability by job type. Information like this can help you decide whether or not to continue to do one type of work. If your commercial jobs aren't as profitable as your residential work, you might decide to stop bidding commercial jobs.

Setting Up a New Job Type

To set up a new job type:

▌ From the drop down **Job Type** list in the Job Info tab window, choose **Add New**.

▌ In the New Job Type window, in **Job Type Name**, enter the new job type name. See Figure 7-7.

▌ If you want the new job type to be a subtype of another job type, click **Subtype of** and enter the appropriate job type. See Figure 7-8.

▌ Click **OK** to enter the new job type and return to the New Job window.

▌ Click **OK** in the New Job window to add the job for the customer.

▌ Click the **X** in the upper right of this window to close the Customer Center.

Exporting or Printing Your Customer List

You can export your Customer:Job list to a program such as Excel, ACT, or any other program that can import delimited text. You can also print out reports for your field personnel.

To run a customer report:

▌ From the **Reports** menu, choose **Customers & Receivables**, then **Customer Contact List.**

▌ In the Modify Report window, select the columns you want to display on your report.

▌ Click **OK**.

To export a customer list:

▌ Create a Contact List report as outlined above.

▌ In the Customer Contact List report window, click **Export**.

▌ Select the export options you wish to use and click **Export**.

Now that you've finished entering your customers and jobs, let's move on to Chapter 8 to set up your vendors and subcontractors.

Chapter 8

Vendors and Subcontractors

In QuickBooks Pro, there's no difference between a subcontractor and a vendor. Whether they supply you with materials, services, or both, you set up vendors and subs the same way. But how you set them up will determine how you can create your reports on them. It's important to closely follow our instructions for setting up vendors, especially when you set up 1099 vendors.

Setting Up 1099 Vendors

You're required to report on IRS Form 1099 the yearly total your company paid each vendor that supplied over $600 of labor. If you send 1099-MISC forms to any vendors or subcontractors, you can set up QuickBooks Pro to track all 1099-related payments to each vendor. Then at the end of the year, QuickBooks Pro can print your 1099-MISC forms.

To set up QuickBooks Pro to track 1099s, you need to set the Tax:1099 preference.

▌ From the **Edit** menu, choose **Preferences**.

▌ Click the **Tax:1099** icon.

▌ In the Company Preferences window, click **Yes** to answer **Do you file 1099-MISC forms?**

Next you need to associate the appropriate accounts with each 1099 category you use to report to the IRS. Most construction businesses report amounts only for Box 7: Nonemployee Compensation.

▌ Click in the Account column next to **Box 7: Nonemployee Compensation**.

▌ In the drop-down Account list, select the account you've been using to track 1099 payments to your vendors. If you've used more than one account, click **Multiple accounts** at the top of the list, and then select each individual account.

Figure 8-1
The Vendor Center window displays each vendor you have entered into QuickBooks Pro and the amount you owe each vendor.

Remember that each 1099 category has to have its own unique account. If you report Nonemployee Compensation (Box 7), you could use your Cost of Goods Sold account named Job Related Costs to track Nonemployee Compensation. But once you select Job Related Costs for Box 7, you can't use it to track any of the other 1099 categories.

▌ Click **OK** to save your preference changes.

Now you're ready to create a new 1099 vendor.

▌ Click the **Vendor Center** icon on the menu bar, or from the **Vendors** menu, choose **Vendor Center**. See Figure 8-1.

▌ In the Vendor Center window, click **New Vendor**. Figures 8-2 and 8-3 show New Vendor windows for Eric Savage Electrical Company.

▌ Enter the vendor's name.

▌ In **Company Name**, enter the name as you want it on your QuickBooks Pro Vendor list. This name won't appear on the 1099 form if you also fill in **First Name**, **M.I.**, and **Last Name**.

If the vendor is a sole proprietor, you have to print the person's name, not the company name, on the 1099. Because of this, you'll need to enter the person's legal name in the First Name, M.I., and Last Name fields. If the vendor's ownership is classified as a partnership, don't fill in the First

Figure 8-2

Use the Address Info tab to add or change a vendor's name, contact, phone, or address.

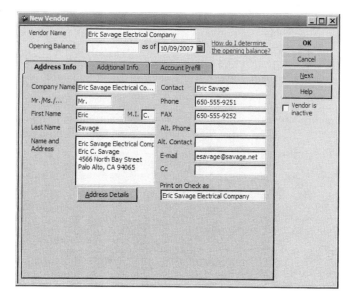

Figure 8-3

Use the Additional Info tab to track more details about a vendor. You can also create custom fields specific to your needs here.

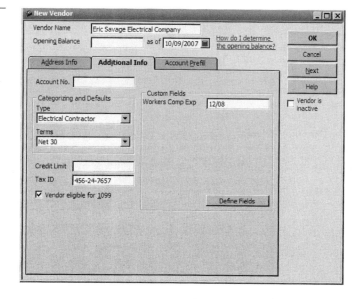

Name, M.I., or Last Name fields (you can enter a contact name in the contact field). Enter the legal partnership name in the Company Name field. If the vendor is incorporated you don't have to send them a 1099.

If part of the address in the Address field includes a name on line 2 or below, be sure the name is identical to the Company Name field or the First Name, M.I., and Last Name fields. Otherwise, it will print as part of the address on the 1099 form.

Figure 8-4

Use the Define Fields window to enter a label for a custom field and specify if the custom field will be used for customers, vendors, or employees.

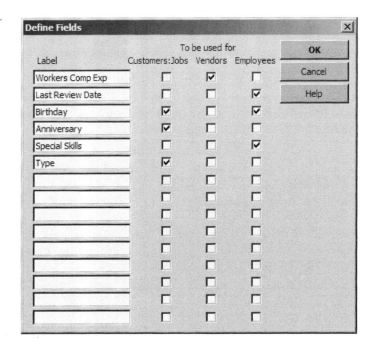

- Be sure you enter the company's two-letter state abbreviation and zip code. Click the **Address Details** button to enter this information.

- Click **OK.**

- Click the **Additional Info** tab. See Figure 8-3.

- Fill in **Tax ID**. If the vendor is a sole proprietor, enter the vendor's Social Security Number or EIN (employer identification number).

- Select **Vendor eligible for 1099**.

- Fill in **Account No.** if you wish. This information will be printed in the Memo field of any checks to the vendor. If you have an account number with a vendor, you can enter it here and it'll automatically be printed on each check to the vendor.

You can set up custom fields in the Additional Info window. One field you might want to add is a custom field called Workers Comp Expires. You can use this field to track the expiration date of the workers' comp insurance for each subcontractor.

To create a custom field:

- In the Custom Fields area of the Additional Info tab window, click **Define Fields**. See Figure 8-3.

- In the Define Fields window, under **Label**, enter the field names you'd like to create. See Figure 8-4.

Figure 8-5

QuickBooks Pro
Contractor edition has
built-in fields for Work
Comp and General
Liability expiration dates.

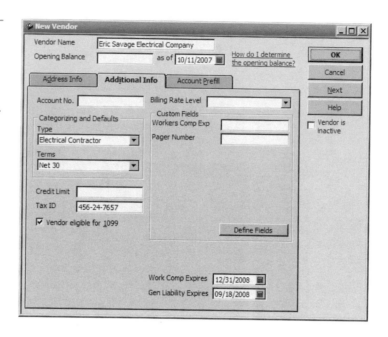

In the columns to the right, check the boxes to select whether the label is used for customers, vendors, or employees.

Click **OK** to create the new labels.

If you're using the QuickBooks Premier Contractor edition, there are two new fields on the Additional Info tab: Work Comp Expires and Gen Liability Expires (Figure 8-5). There's no need to create custom expiration dates if you're using the Contractor Edition. Simply enter the expiration dates and, when you enter transactions, you'll be notified if the insurance expiration dates have passed.

The Account Prefill tab is only for use with non-1099 vendors (see next section). Disregard this tab when you're entering 1099 vendors.

Now you're ready to save all information about the vendor:

Click **OK**.

Setting Up a Non-1099 Vendor

You set up a non-1099 vendor the same as a 1099 vendor, with these two exceptions: First, you don't have to enter a tax ID number. Second, you don't select Vendor eligible for 1099 in the New Vendor Additional Info tab window. For Vendor Type, enter the appropriate type such as HVAC contractor, electrical subcontractor, and so on.

You can opt to "prefill" expense accounts when entering transactions. This adds speed and accuracy to your data entry.

Figure 8-6
Use the Account Prefill tab only for non-1099 vendors.

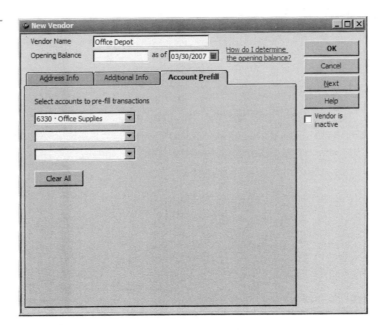

Click the Account Prefill tab (Figure 8-6). Here, you can choose up to three expense accounts per vendor, that will automatically appear when you enter vendor bills and write checks.

Now that you've finished setting up your vendors, let's set up your employees. If you don't have employees, skip Chapter 9 and move on to Chapter 10 where we will cover Opening Balances.

Chapter 9

Employees

There are two cases when you'll want to set up your employees in QuickBooks Pro using the Employee Center. First, use it if your company has employees and you process your own payroll. Second, use it if you want to track employee time by job even though your payroll is processed by a service.

Setting Up Your Employee List

The Employee Center includes a listing of every employee record you enter into QuickBooks Pro. Each employee's record contains all the information you need to calculate payroll, whether the employee is on monthly salary or getting a weekly paycheck based on hours worked.

The Payroll module in QuickBooks Pro is very flexible. You can use it to figure out all common payroll items such as federal taxes, FUTA, Social Security, Medicare, state unemployment insurance, and state disability. Naturally, you can track, calculate, and print W2s, 1099s, and Forms 940 and 941. You can also track sick time and vacation pay and create your own custom additions and deductions.

It's easy to add, edit, or delete employees from the Employee Center. To add employees, use the employee template you set up in the Payroll and Employees preference in the Employee Defaults window. The template contains the information common to all employees, so you only have to enter individual employee information to set up a new employee. If an employee goes out on disability or maternity leave, you can assign him/her as "Inactive."

To set up an employee:

▌ Click the **Employee Center** icon in the menu bar, or from the **Employee** menu, choose **Employee Center**. See Figure 9-1.

Figure 9-1
The Employee Center window displays the Social Security number and the name of every employee you have entered into QuickBooks Pro.

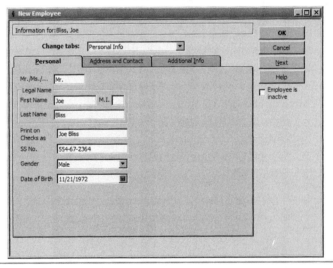

Figure 9-2
Use the Personal tab to add or change an employee's name, Social Security number, etc.

▌ In the Employee Center window, click **New Employee**. The figures in this chapter show how we entered new employee Joe Bliss. See Figure 9-2. You can use them as an example for entering your employees.

Using the Personal Tab

▮ Fill in **Mr./Ms.**, **First Name**, **M.I.**, and **Last Name**.

▮ In **Print on Checks as**, enter the name as you want it to appear on the employee's paycheck and W2.

▮ In **SS No.**, be sure you enter the employee's Social Security number correctly. It's used on quarterly reports and W2s.

▮ The **Gender** and **Date of Birth** fields are for reference only. You use them to store employee information.

Using the Address and Contact Tab

▮ Be sure you enter the employee's address, city, two-letter state abbreviation and zip code.

▮ **Phone**, **Cellular**, **Alt. Phone**, **Fax**, **E-mail** and **Pager** fields are also for reference only. Use them to store employee information.

Using the Additional Info Tab

Use **Custom Fields** to create new fields to collect additional data for all your employee records. QuickBooks Pro treats the information you enter in a custom field the same way it treats information entered into any other field. If you export a list that contains data in custom fields, QuickBooks Pro exports that data along with the other data from the list. Then you can sort the exported field and use it in reports that QuickBooks Pro doesn't provide.

Here are examples of some employee custom fields you can use:

▮ last review date

▮ birthday

▮ pager number

▮ productivity numbers

▮ goals

▮ special skills

To create a custom field:

▮ In the New Employee window, click the **Additional Info** tab. See Figure 9-3.

▮ In the Custom Fields area, click **Define Fields**. See Figure 9-3.

▮ In the Define Fields window, under **Label**, enter the field names you want to create. See Figure 9-4.

▮ In the columns to the right, check the box for **Employees**.

▮ Click **OK** to create the new fields.

Figure 9-3
Use the Additional Info tab to track more details about an employee. Create custom fields you need here.

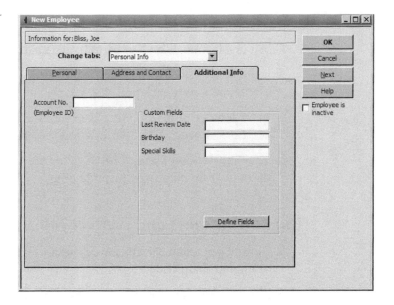

Figure 9-4
Use the Define Fields window to enter a label for a custom field and specify if it will be used for customers, vendors, or employees.

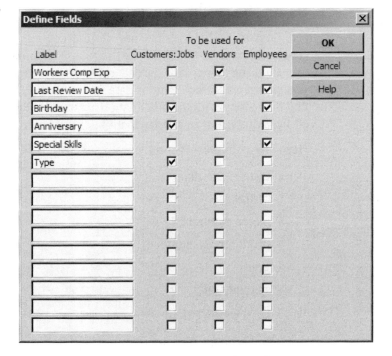

Using the Payroll Info Tab

Using the Payroll Info tab window, you can set up information for each employee on:

- earnings

Figure 9-5

Use the Payroll Info tab to record earning information, pay period frequency, additions, deductions, company contributions, and tax-related information.

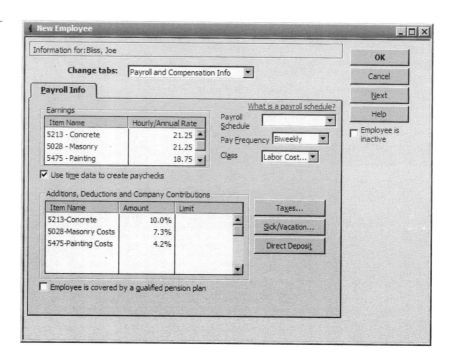

additions, deductions, and company contributions

pay period

class

taxes

sick/vacation

direct deposit

To get to the Payroll Info tab window:

In the New Employee window, from the **Change tabs** drop-down list, select **Payroll and Compensation Info**. See Figure 9-5.

Under **Earnings**, select all payroll items from the drop-down list that apply to the employee. (See Chapter 5 if you haven't set up payroll items.) For example, the employee in our example works on concrete and/or masonry and painting so his payroll information includes the concrete, masonry and painting payroll items. The payroll item determines what account the payroll expense is posted to, such as to 5110 - Job Related Costs.

Select an **Hourly/Annual Rate** for the employee. Realize that the payroll item determines where the payroll expense is posted, and the hourly/annual rates are the actual base wage being used in calculating the payroll as well as the job costing. For **Hourly**, enter in the hourly rate. For **Annual Rate**, enter annual salary.

Figure 9-6

Use the Federal, State and Other tabs to specify tax information for each employee.

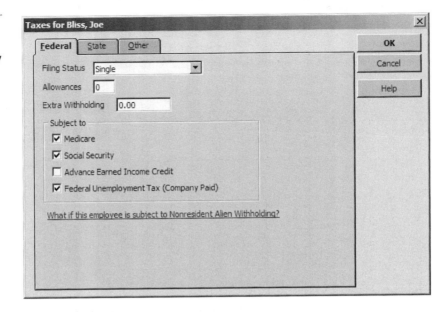

- Select the **Use time data to create paychecks** box if you plan on using the QuickBooks Pro timecard to enter the employee's time. If you select this button, the timecard information entered will flow into the paycheck for this employee.

- Under Additions, Deductions, and Company Contributions, select all payroll items that apply to the employee.

- Select the appropriate **Pay Period** for the employee.

- Click **Taxes** and make sure you select all the appropriate boxes. See the example in Figure 9-6. Click **OK**.

- Click **Sick/Vacation** if the employee is eligible for sick or vacation pay. See the example in Figure 9-7. Click **OK**.

- If you've subscribed to QuickBooks Pro's payroll service, click **Direct Deposit** to set up direct deposit of the employee's pay if appropriate, as seen in Figure 9-8. Click **OK**.

Now you're ready to save all information about the employee:

- In the New Employee window, click **OK**.

Setting up a Sole Proprietor/Partner (Employee) Card

In Chapter 5 we discussed how to include a sole proprietor or partner's time in job cost reports. In this section, we'll walk you through setting up the employee card for the sole proprietor or partner. Keep in mind that a sole proprietor or partner (non-incorporated business) does not get paid through payroll; instead, they take money out of the business in the form of a with-

Figure 9-7

For eligible employees, use the Sick & Vacation window to track available hours and accrual information.

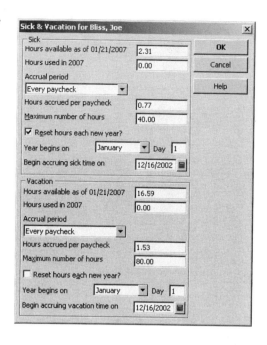

drawal. You will set up an employee card if the owner works on jobs and wants to track the costs. In Chapter 15 we'll walk you through processing a zero payroll check to the owner (in order to job cost the owner's time to a job).

To set up a sole proprietor/partner employee card:

▌ In the **Employee Center** window, click **New Employee**. Refer back to Figure 9-1.

▌ Enter the Employee information on the Personal, Address and Contact, and Additional Info tabs. Refer to Figures 9-2 and 9-3.

▌ In **Change tabs**, select **Payroll and Compensation Info**. See Figure 9-8.

▌ In the **Earnings** box, select **Owner's Labor** (if you don't see it in the list refer back to Chapter 5, Payroll Items).

▌ In **Hourly/Annual** rate, enter the amount you would pay someone else with the same qualifications to do the work the owner performs on the job. The rate should be per hour including labor burden. Labor burden is the cost of payroll taxes, workers' comp and employee benefits. Do not enter in the amount the owner withdraws; instead you should enter the amount you would have to pay someone else to do that job. Later, when you are analyzing your job costs reports, you'll want to know which jobs are profitable, including all the time the owner put into the job.

Figure 9-8

Set up payroll information to record job costs for a sole proprietor or partner.

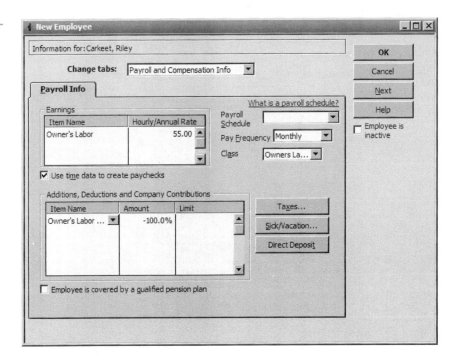

▌ In **Class**, select **Owner's Labor**. If it doesn't appear in your list, you should add it now. See Chapter 6 for more information about classes.

▌ Click the button next to **Use time data to create paychecks**.

▌ Under Additions, Deductions and Company Contributions, enter **Owner's Labor Deduction** (if it doesn't appear in your list, see Chapter 5 to learn how to enter it). In **Amount**, enter 100%. It will automatically appear as negative 100% (–100%). You'll learn more about why this is important in Chapter 15.

▌ Click **OK** to save the employee information.

Now that you've finished setting up customers, jobs, vendors, and employees, it's time to tell QuickBooks Pro what your opening balance is for each account you have. Chapter 10, Opening Balances, will show you how to enter those balances.

Chapter 10

Opening Balances

Getting your computerized accounting system into balance can be a trying experience. If you're already using a computerized accounting system, you should make sure your balances are accurate before transferring them over to QuickBooks Pro.

Keep in mind when you begin using QuickBooks Pro that you don't have to start with opening balances and can always go back and change your beginning balances.

Entering Opening Balances in QuickBooks Pro

A journal entry is used to get your beginning balances into QuickBooks. If you don't know what your beginning balances are, your accountant or tax preparer can help you arrive at those numbers.

▌ From the **Company** menu, choose **Make General Journal Entries** to get the General Journal entry window.

▌ Enter **Asset**, **Liability**, and **Capital** account balances. Keep in mind that Assets usually have debit balances (except accumulated depreciation), while Liabilities and Capital have credit balances. See Figure 10-1.

> **Note:** *If you're starting to use QuickBooks Pro at the beginning of your fiscal year, you enter only Asset, Liability, and Capital account balances. Don't enter beginning income, cost of goods sold, or expense account balances.*

Figure 10-1

This is an example journal entry to get beginning balances into QuickBooks.

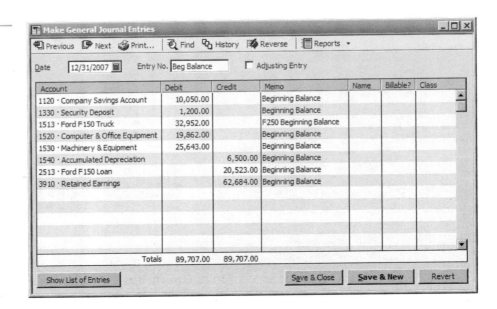

When you finish, print a Balance Sheet to make sure everything balances. To print a Balance Sheet:

▌ Click **Home**, then click **Chart of Accounts**, or from the **Lists** menu, choose **Chart of Accounts**.

▌ In the Chart of Accounts window, pull down the **Reports** menu, choose **Report on All Accounts**, then **Balance Sheet**, and then **Standard**. The date of the report should be the date you started using QuickBooks Pro.

> **Note:** Don't enter Accounts Receivable or Accounts Payable balances. They'll come in automatically when you enter the outstanding invoices (receivables) and outstanding bills (payables).

Entering Invoices for Accounts Receivable

Accounts Receivable is money your customers owe you as of the date you start using QuickBooks Pro to enter transactions. If you've been using another accounting system, you should be able to print out a report with each customer's name, date of each outstanding invoice, invoice number, and the amount due and transfer the information to QuickBooks Pro manually.

Figure 10-2

Use the Create Invoices window to enter each outstanding invoice to make sure your Accounts Receivable balance is correct.

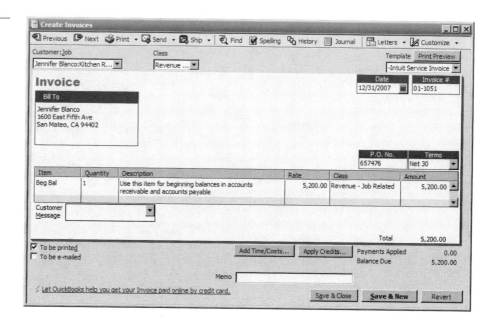

Enter each outstanding Accounts Receivable invoice as an Intuit Service Invoice with the date of the original invoice, *not* the date you enter the invoice. You won't be sending out these invoices — they're simply a way to enter your Accounts Receivable balances. In Figure 10-2, we've entered a beginning balance for Jennifer Blanco.

To enter an outstanding customer invoice:

▌ Click **Home**, then click **Create Invoices**, or from the **Customers** menu, choose **Create Invoices**.

▌ In **Template**, select **Intuit Service Invoice** from the drop-down list.

▌ In **Customer:Job**, select the customer's name and job. If you entered an estimate for this job, you'll see the Available Estimates dialog box. Click **Cancel** to close the dialog box.

▌ In **Date**, enter the original date of the invoice, not the current date.

▌ In **Invoice #**, enter the original invoice number.

▌ In **P.O. No.**, enter the purchase order number for the invoice or job if the customer gave you one.

▌ In **Terms**, select the appropriate terms for the invoice from the drop-down list.

▌ Click in the **Item** column.

Figure 10-3

Use the Enter Bills window to enter each outstanding bill to make sure your Accounts Payable balance is correct.

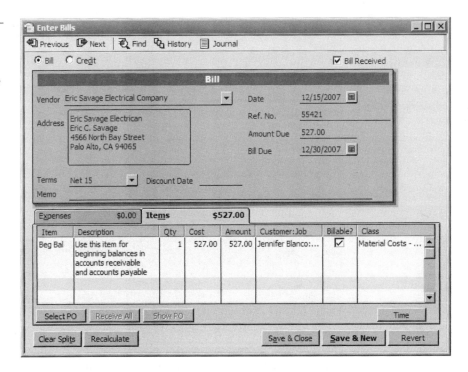

- From the pull-down Item list, select **Beg Bal**.

- In **Amount**, enter the amount of the invoice.

- Click **Save & Close**.

Entering Bills for Accounts Payable

Accounts Payable is money you owe to vendors (and subcontractors) as of the date you start entering transactions into QuickBooks Pro. If you've been using another accounting system, you should be able to print out a report with the vendor's name, date of each outstanding bill, bill number, and the amount due and transfer the information to QuickBooks Pro. Enter each outstanding Accounts Payable bill with the date of the original bill, *not* the date you enter the bill. In Figure 10-3, we've entered an outstanding bill from Eric Savage Electrical Company.

To enter vendor outstanding bills:

- Click **Home**, then click the **Enter Bills** icon, or from the **Vendors** menu, choose **Enter Bills**.

- In **Vendor**, from the pull-down list, select the vendor's name.

- In **Date**, enter the date of the bill, *not* the current date.

▌ In **Ref. No.**, enter the vendor's invoice/bill number.

▌ In **Amount Due**, enter the full amount of the bill (including tax, shipping, etc.).

▌ In **Terms**, select the appropriate terms for the bill from the drop-down list.

▌ Enter a memo for future reference.

▌ Click the **Items** tab.

▌ Click in the **Item** column.

▌ From the Item pull-down list, select **Beg Bal.** Enter **Qty**, **Cost**, **Amount**, **Customer:Job**, and **Class**.

▌ If the amount is going to be billed to your customer, check the box in the Billable column.

▌ Click **Save & Close**.

Now you've set up opening balances in QuickBooks Pro and you should be ready to enter new transactions in your company's data file. Remember though, you can always go back and make changes to the file if you find that something is out of balance.

Chapter 11

Organizing Work Flow

Even with the latest computer hardware and all the power of QuickBooks Pro, you still may not have all the data you need. Why? Because to get the most from your system, you need to have an organized flow of information from the field, through the office, to your computer system. For QuickBooks Pro to do its job you'll need to study — and possibly change — how work flows at your business.

For example, QuickBooks Pro has a flexible and powerful payroll capability. It's especially useful for allocating the cost of labor to multiple jobs. But unless you get that payroll information into your computer system, you won't be any better off than before you started using QuickBooks Pro.

Whether you work alone or have office help, office and paperwork flow should follow a strict process. If you don't have one consistent written method for organizing paperwork, you may find yourself wasting many valuable hours shuffling papers rather than generating income for your business.

The best way to get a new routine organized is to get everyone involved. For example, before you start using a new payroll timecard or a new process for turning in paperwork, plan a lunch meeting with all your field workers. During lunch give them something new that will help them stay organized, such as a binder with their new timecards in it, or a clipboard to keep receipts together until they get back to the office. Treating them to lunch and providing free organizing materials may help motivate them to follow your new procedures.

In our opinion, the more informed you keep the people actually doing the work, the more effective they can be. Usually job cost information needs to make a complete circle so the people creating it get feedback on how they're doing. We recommend that you show your project managers a job cost report once a week. If they understand how you're using the data they provide, they'll be more likely to make sure you get it.

The Job Estimates vs. Actuals Detail is a good report to give your project manager on a weekly basis. It shows a job, broken out by phase, with its estimate, actual, and variance figures. These numbers will help the project manager see where his job is running over, or under, the estimates — while he can still do something about it. Another useful report for project managers is the Vendor Contact List Report. This report includes company name, contact name, phone number, and fax number.

Setting Up Your Office Files

One of the hardest jobs in a construction office is keeping the files current. You need an effective filing system to keep things from piling up. Here's the system we recommend.

For your financial files, let QuickBooks Pro be your guide. Organize your file drawers the same way QuickBooks Pro organizes its lists — by vendors, employees, and so on. It makes good sense to take advantage of the correlation between the QuickBooks Pro lists and your paper files. Intuit spent many hours and dollars figuring out the most efficient system for filing with a computerized system. Put their expertise to work for you, and use this same system for your paper files.

Customers and Jobs Files

You'll need a file drawer, or a section of a file drawer, for customers, with separate folders for each job that you do for that customer. For example, Rachel Olsen originally contacted us about remodeling her kitchen, and at Twice Right we have one folder labeled "Olsen, Rachel — Kitchen Remodel." Then when she decided to build a guest cottage, we started another file labeled "Olsen, Rachel — Guest Cottage." We filed all correspondence to and from Rachel in one of those folders.

Your customer folder should contain:

- any correspondence to or from the customer about the job.
- the original project contract.
- copies of the billings. If the billings are bulky, use a separate customer job file for billing information.
- any change orders you issue.

Your customer folder should *not* contain:

- vendor invoices for the job.
- any documents that don't relate to the job.

Figures 11-1 and 11-2 are flow charts showing recommended job and customer transaction procedures.

Job Transaction

Figure 11-1

This flow chart shows how a job is processed. To see your jobs in QuickBooks Pro, click the Customer Center icon.

Customer Transaction

Figure 11-2

This flow chart shows how things you enter in QuickBooks Pro affect a customer.

Vendor Files

You'll need a file drawer, or a section of a file drawer, for vendors. But it's really not necessary to give all vendors their own files. If you only get a few bills a year from a vendor, you may want to file that vendor's invoices in a Miscellaneous Vendors file. When the Miscellaneous Vendors file gets too big, break it down into two files, such as Miscellaneous Vendors A-L and Miscellaneous Vendors M-Z. File all bills in the vendor files whether they're job related or not.

You'll want to set up an individual file for some vendors. For example, at Twice Right Construction we have a charge account with Truitt & White Lumber Company. Because we get several invoices a month from Truitt & White, Truitt & White has its own file.

The chart in Figure 11-3 shows how to organize vendor transactions.

Figure 11-3

This flow chart follows the path of a vendor bill from the time it's received until the bill is paid.

Vendor Transaction

Employee Files

You'll also want to keep a file drawer, or a section of a file drawer, for employees. Since this information is confidential, be sure to keep these files in a locked file cabinet.

Your employee folder should contain:

- a completed employment application.

- a completed I-9.

- the employee's resume, if you have one.

- employee review documents.

- benefits documents.

Organizing Your Payroll

One way to make sure you report labor accurately, on a per job basis, is to use items that describe the phases of labor. See Chapter 4 if you need more information on these items. You can print these items on the back of time-cards so employees can use them when they fill out the timecards. Show your employees how using the items helps the business determine true job costs and helps raise profits (and wages). They'll be more likely to fill them out carefully if they understand their importance.

Once the timecards are filled out, it's a simple matter of entering the items into a timesheet in QuickBooks Pro. If you make the items on the back of the timecard match the names of the items in QuickBooks Pro, then entering the timesheet in QuickBooks Pro will simply be a matter of selecting the matching numbers. For example, suppose Joe Bliss has completed 40 hours of carpentry work in a week. His paper timecard would look something like Figure 11-4. In QuickBooks Pro, the timecard would look like Figure 11-5. The "13" item in the Task # column of the timecard matches the "13 Windows" item in the Service Item column on the Timesheet.

If you use computers at your job sites, you can fill in electronic time-cards or summary sheets and e-mail them as attachments to the home office.

Keeping Office Paperwork Current

The most important thing is getting the field paperwork into the office. You'll have to start by working with your field people on how, and when, you want them to turn paperwork into the office. Don't get discouraged. When you first start out, you'll probably get all the packing slips, invoices,

	Date:	1/5/2007										
	Employee:	Joe Bliss										
	Pay period ending:	1/7/2007										

Customer/Job	Task #	Description	Mon	Tue	Wed	Thur	Fri	Sat	Sun	Total
Blanco/Kitchen	13	Windows	8	8	8					24
Blanco/Kitchen	21	Cabinets				8	8			16
Daily totals:			8	8	8	8	8			40

Figure 11-4

A typical paper timecard computed for a week.

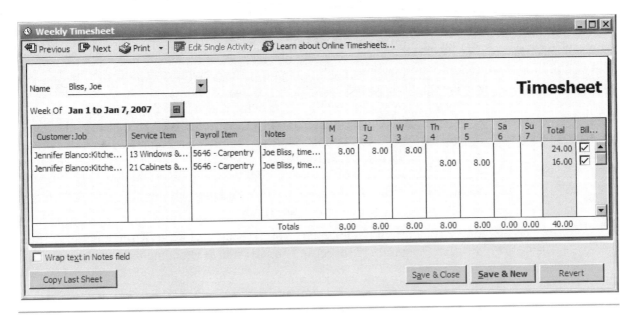

Figure 11-5

A QuickBooks Pro timesheet looks similar to a paper timecard. Notice how the job and item entered match the job and item on the paper timecard.

and receipts turned in torn and crumpled, or not turned in at all, with excuses for why they were lost. But with encouragement, your field people will get the hang of it and start turning in their paperwork on a regular basis.

You'll need a place in the office, such as a basket or box, to use as an inbox for all field and office paperwork that comes into the office. Make sure the field staff can see and reach this inbox easily when they come in to drop off their paperwork.

Sort the paperwork that finds its way to the office. Put payroll timecards in a "To be approved" file. Put vendor invoices in a file for coding and approval.

Enter approved timecards and vendor invoices in QuickBooks Pro. Then move them to the "To be paid 1 - 15th" file or the "To be paid 16th - EOM" file. After you pay an invoice or timecard, file it in the vendor file or employee file.

Another handy job-costing tip is to ask your materials suppliers and subcontractors to bill you by job. Most suppliers won't have any trouble issuing a separate invoice (or at least a separate line item) with a specific job name. Usually you can specify that no materials can be purchased without a job name being assigned to the invoice. As a check, you can also ask your field personnel to save shipping slips and bills of lading and to write the job name on them.

Change Orders

Whether you issue change orders in the office or the field, you need to set up a process to handle them consistently and smoothly. It's important that you inform the office as soon as possible when there's a change order. Change orders can have a major impact on the profitability of a job. The sooner you get them into your system and begin job costing, the sooner you'll know the impact of the changes.

In the first eleven chapters of this book, we concentrated on showing you how to set up QuickBooks Pro for a construction business. In the next chapter, we'll show you how to enter an estimate in QuickBooks Pro.

Chapter 12

Estimating

In this chapter we'll show you how to bring outside summary estimate totals into QuickBooks Pro. We realize that estimating is an individual thing. If your style of estimating is working for you, keep using it. Or if you already use, and are happy with, a sophisticated estimating program that returns useful information that QuickBooks Pro doesn't track, stick with it. You can use a spreadsheet program, or Craftsman's National Estimator, or any of the other estimating programs on the market. As long as your estimating program can create a summary by phase report, you can use it to generate an estimate to hand-enter into QuickBooks Pro.

You can enter the summary information into QuickBooks Pro and track it against your actual costs. Then you can generate a report called Job Estimates vs. Actuals Detail. This report is the most useful report you can get for running your jobs on a day-to-day basis.

QuickBooks Pro has no labor and material cost database. If you use 100 or fewer different items, you may be perfectly happy to enter those 100 items into the QuickBooks Pro Item list. When your costs change, you can go back to the Item list and change your selling prices. But if you handle a wide variety of work and want to be ready to estimate almost anything, or if you use many unique items grouped together in a couple of dozen phases to make your estimates, you'll be asking too much from QuickBooks Pro.

In this chapter, we'll just show you how to enter a summary estimate into QuickBooks Pro.

Name	Description	Type	Account	On Hand	Price
◆ 01 Plans & Permits	Plans & Permits	Service	4110 · Construction Income		0.00
◆ 01.1 Plans	Plans	Service	4110 · Construction Income		0.00
◆ 01.2 Building Permits	Building Permit	Service	4110 · Construction Income		0.00
◆ 01.3 City & Co. Lic's & Fees	City & County Licenses & Fees	Service	4110 · Construction Income		0.00
◆ 02 Site Work	Site Work	Service	4110 · Construction Income		0.00
◆ 02.10 Demo	Demolition	Service	4110 · Construction Income		0.00
◆ 03 Excavation	Excavation	Service	4110 · Construction Income		0.00
◆ 04 Concrete	Concrete	Service	4110 · Construction Income		0.00
◆ 05 Masonry	Masonry	Service	4110 · Construction Income		0.00
◆ 06 Floor Framing	Floor Framing	Service	4110 · Construction Income		0.00
◆ 07 Wall Framing	Wall Framing	Service	4110 · Construction Income		0.00
◆ 08 Roof Framing	Roof Framing	Service	4110 · Construction Income		0.00
◆ 09 Roof Flashing	Roof Flashing	Service	4110 · Construction Income		0.00
◆ 10 Exterior Trim & Decks	Exterior Trim & Decks	Service	4110 · Construction Income		0.00
◆ 11 Siding	Siding	Service	4110 · Construction Income		0.00
◆ 12 Doors & Trim	Doors & Trim	Service	4110 · Construction Income		0.00
◆ 13 Windows & Trim	Windows & Trim	Service	4110 · Construction Income		0.00
◆ 14 Plumbing	Plumbing	Service	4110 · Construction Income		0.00
◆ 15 HVAC	Heating & Cooling	Service	4110 · Construction Income		0.00
◆ 16 Electrical & Lighting	Electrical & Lighting	Service	4110 · Construction Income		0.00
◆ 17 Insulation	Insulation	Service	4110 · Construction Income		0.00
◆ 18 Interior Walls	Interior Walls	Service	4110 · Construction Income		0.00
◆ 19 Ceilings & Cover	Ceilings & Coverings	Service	4110 · Construction Income		0.00
◆ 20 Millwork & Trim	Millwork & Trim	Service	4110 · Construction Income		0.00
◆ 21 Cabinets & Vanities	Cabinets & Vanities	Service	4110 · Construction Income		0.00
◆ 22 Specialty	Specialties	Service	4110 · Construction Income		0.00
◆ 23 Floor Coverings	Floor Coverings	Service	4110 · Construction Income		0.00
◆ 24 Paint	Painting	Service	4110 · Construction Income		0.00
◆ 25 Cleanup	Cleanup & Restoration	Service	4110 · Construction Income		0.00
◆ 26 Landscape & Paving	Landscape & Paving	Service	4110 · Construction Income		0.00
◆ 27 Contingency	Contingency	Service	4110 · Construction Income		0.00

Item ▾   Activities ▾   Reports ▾   Excel ▾   ☐ Include inactive

Figure 12-1
The Item List in *sample.qbw*.

Using a Summary Estimate You Make Outside of QuickBooks Pro

Before you transfer a summary estimate, it's a good idea to set up an item in QuickBooks Pro for each phase you use in your estimating program. See Chapter 4 for information on how to set up items. For example, suppose you decide to use our job phases partially shown in Figure 12-1. Then in QuickBooks Pro, you should set up these items. If you used our *company.qbw* data file to set up your own company data file, these items are already set up for you.

You can also create subitems to get more detail on the phases. For example, you could set up an item for the Plans and Permits phase with subitems for plans, building permits, and city license. Subcontractors may need an entirely different list of items — one that divides their work into appropriate categories.

Once you've entered the items (or modified existing items), you can manually transfer the information from each phase in a summary estimate to a matching QuickBooks Pro item. Let's walk through entering an estimate in QuickBooks Pro.

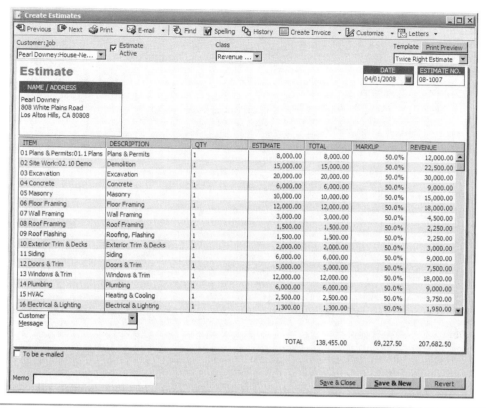

Figure 12-2

Use the Create Estimates window to quickly and easily create a new estimate for a customer.

To create a new estimate in QuickBooks Pro using summary figures from an outside estimating program:

▌ Click **Home**, then click the **Estimates** icon, or from the **Customers** menu, choose **Create Estimates**.

▌ From the drop-down **Customer:Job** list, select the customer. In our example shown in Figure 12-2, we selected Pearl Downey:House — New Construction.

▌ From the pulldown class list, select Revenue - Job Related.

▌ Enter **Date** and **Estimate No**.

▌ To enter each line item for the estimate, click in the **Item** column, and select the item that matches the item in the outside summary estimate from the pull-down list. The description of the item will fill in automatically if you entered it when you created the item.

▌ In **Qty** and **Cost**, enter the figures for the item from the outside summary estimate. Amount will be calculated for you by multiplying Qty by Cost. The estimate form in Figure 12-2 has been customized to show columns for Qty, Estimate, Total, Markup, and Revenue,

Figure 12-3
The Additional
Customization window
allows you to customize
the estimate to fit your
company's needs.

instead of Qty, Cost, Amount, Markup, and Total. We'll walk you
through customizing an estimate form in the next section.

▌ In **Markup**, enter the number to add a lump sum markup. To add a
percentage, enter the number and a percentage sign (say 50%).
QuickBooks Pro will calculate the appropriate markup, add it to the
Amount, and show the result in Total.

▌ Click **Save & Close** to create the estimate.

Customizing an Estimate Form

You can change the way an estimate looks both on paper and on your computer screen. To do this:

▌ Click **Home**, then click the **Estimates** icon, or from the **Customers**
menu, choose **Create Estimates**.

▌ Click on **Customize**.

▌ Click on **Manage Templates**. Select an estimate template and click
Copy.

▌ In **Template Name**, enter a name for your customized estimate.
Click **OK**.

▌ Click on **Additional Customization.** See Figure 12-3.

Figure 12-4

Memorize estimates that contain line items you use frequently. Be sure to give the memorized estimate a distinctive name, such as Kitchen Remodel, so it's easy to find when you need it.

Take a look at each of the four tabs in the Additional Customization window. You can make changes to the form, both in what appears on the screen and what is printed. You can change field labels, add or remove columns, create page footers, and set printer parameters.

▌ When you're done, click **OK**.

Back in the Basic Customization window, you can add your company logo, or any picture, to the top of your estimate by selecting the **Use Logo** box. Click on the **Layout Designer** to change the estimate column widths.

Memorizing an Estimate

It's a good idea to memorize an estimate that has line items you use frequently. For example, you could create an estimate for a kitchen addition, and memorize it as Basic, Medium, or High-end Kitchen. Then the next time you have a kitchen addition job to estimate, simply open the memorized estimate that's most like your new job and make changes to the items that are different.

Once you memorize an estimate, you can use it as a template — like a cookie cutter to stamp out more estimates with the same items, but with different customer or job names. This can save you many hours of estimating time. There's no need to start from scratch every time you estimate a job.

To memorize an estimate:

▌ With the estimate you want to memorize on your screen, from the **Edit** menu, choose **Memorize Estimate** and click **OK**.

▌ In the Memorize Transaction window, enter a name for the estimate. In our example shown in Figure 12-4, we entered *Kitchen Remodel*.

▌ Click **Don't Remind Me** because there's no need to be reminded regularly of a memorized estimate.

▌ Click **OK** to memorize the estimate.

▌ Click the **X** in the upper right of the window to close the Estimate screen.

To use a memorized estimate:

▌ From the **Lists** menu, choose **Memorized Transaction List**.

▌ In the Memorized Transaction List window, select the memorized estimate you want to use.

▌ Click **Enter Transaction**.

Now you'll see a Create Estimates window with nothing in Customer:Job and Name/Address.

▌ From the pull-down Customer:Job list, select the customer. Or select **Add New** to add a new customer. For more information on adding customers, see Chapter 7, Customers.

Of course, you'll probably never have two identical estimates for two customers. Usually, you'll have to change at least a few things in a memorized estimate. You'll probably need to change the quantities and costs, and add or delete lines.

To add (or delete) a line to an estimate:

▌ Click in the **Item** column.

▌ From the **Edit** menu, choose **Insert Line** (or **Delete Line**) and make the necessary changes.

▌ Click **Save & Close** and you've got a new version of an old estimate. Note that the memorized estimate hasn't changed at all. You can still use it for another estimate.

QuickBooks Pro lets you enter multiple estimates for one job. The advantage of having multiple estimates is that you can easily track the changes to the original estimate and proposed contract amount. Then, as the scope of work changes, you can compare the original estimate to the existing estimate and review it with the customer.

To enter a revised estimate, you should first create an estimate as described in the beginning of this chapter. Then, with the original estimate on your screen:

▌ From the **Edit** menu, select **Duplicate Estimate**.

▌ When the estimate is duplicated, make sure you change the **ESTIMATE NO.** field to reflect the revision number. For example, if your original estimate number was 01-1001, your revised estimate should be entered as 01-1001R1. If you have another revision, you should duplicate the latest estimate (01-1001R1 in our example) and enter R2 (or 01-1001R2) in the **ESTIMATE NO.** field.

▌ Click **Save & Close** to create the revised estimate.

Estimates and Progress Billing

After you create an estimate, you can easily create an invoice based on the estimate, using a progress billing invoice. For information on how to create a progress invoice, see Chapter 13. Before you create the invoice, you can see a summary of what you've billed to date, the current charges, and what will be billed later — all by phase of construction (or the items you've set up). So your final invoice is sure to cover everything in the job you haven't billed previously. You can make changes in the estimate any time you want.

From an estimate, you can:

1. Bill 100% — everything that remains to be billed.

2. Bill certain items only. You select which phases you want to bill.

3. Designate a percentage to be billed on any item you select.

If you handle larger, complex jobs that require multiple payments, progress billing may be a godsend. But progress billing only works on estimates, not invoices. To do progress billing you have to start with an estimate.

That's about all there is to know about bringing summary estimates into QuickBooks Pro.

Getting Detailed Estimates

There are ways you can use QuickBooks Pro to produce detailed estimates. One way involves creating a separate Items list in QuickBooks Pro — a somewhat complex task that requires more knowledge of QuickBooks Pro than we cover in this book. We wrote this book to take you step-by-step through a simple way to set up your bookkeeping in QuickBooks Pro. If you're comfortable with QuickBooks Pro and our basic setup, and are willing to give it a try, look in Appendix A. You'll find full instructions on setting up and using detailed cost estimates.

A far better option is Craftsman's Job Cost Wizard. Using this software, you can make estimates using Craftsman's estimating program, National Estimator, and import them directly into QuickBooks Pro. From there, you can track job costs for each item or category in an estimate, compare actual with estimated costs for each part of any job, and create and send invoices. Job Cost Wizard, National Estimator, and a 200-page cost estimating database are included on the CD in the back of this book. Appendix B gives detailed step-by-step instructions on how to use the National Estimator program to estimate and track job costs.

Receivables

In this chapter, and the next, we'll show you how to use QuickBooks Pro on an accrual basis. The accrual basis of accounting means that you track how much money is owed to you (Accounts Receivable) and how much money you owe (Accounts Payable).

The accrual method has both benefits and drawbacks. Some of its benefits are:

▌ you can easily find out how much money you billed to customers vs. how much it cost you to earn that income

▌ you can easily find out how profitable your business was during a particular time period

▌ your job cost reports will include all costs and invoices to date

The major drawback to the accrual method is that it takes time to enter the transactions (bills and invoices). You have to enter every transaction as you receive or generate it. This requires vigilance to keep up to date. If you're a small contractor with little or no help, or no time to do bookkeeping, you're probably operating your business on a cash basis. If so, you can skip to Chapter 15, or keep reading to prepare for the time when you might switch to the accrual method.

Now on to the subject of this chapter — managing your receivables using the accrual system of accounting. Of course, you won't have any receivables if you don't send out bills, and contractors don't like to send out bills. However, with QuickBooks Pro it's so simple there's no excuse for having a backlog of invoices that you need to send out. When you've completed a job, or at designated stages during the job, you'll want to prepare invoices to submit to your customer for payment.

In this chapter we'll show you:

▌ four different ways to invoice a customer

▌ how to handle change orders

   ▌ how to handle retainage

   ▌ how to record customer payments

   ▌ how to enter a deposit

   ▌ how to record job deposits (customer advances)

Four Ways to Invoice a Customer

The way you prepare an invoice will depend on the terms of your contract. The main types of contracts are:

   ▌ set price (usually a small and simple verbal contract)

   ▌ time and materials or cost plus

   ▌ fixed price (usually a contract requiring a certain amount be paid when a certain percentage of the work has been completed)

   ▌ progress or AIAG702 structure

Creating a Set Price Invoice

Sometimes you need to get an invoice out the door quickly. This is usually for non-time-and-materials billings — perhaps a small, short-term job that doesn't require an estimate. For example, suppose a customer phones you to come out and fix a hole in a wall and you tell her you'll come and fix it today. Since she's a good customer who frequently refers your company to others, you agree to do the job for a flat $250. In this case, there's no estimate and the invoice is billed based on a set price.

Here's how you would enter a set price invoice:

   ▌ Click **Home**, then click **Create Invoices**, or from the **Customers** menu, choose **Create Invoices**.

   ▌ From the pull-down **Customer:Job** list, select the customer. The Bill To section will fill in automatically if you completed that part of the customer record. See our example in Figure 13-1.

   ▌ Enter **P.O. No.** and any other general information you need.

   ▌ From the pull-down **Class** list, select **Revenue - Job Related**.

   ▌ Click in the **Item** column and select an item from the drop-down list. Keep in mind that your items represent job phases. **Description** will fill in automatically if you entered it when you set up the item. Change description to fit the job. For this example, we used item 18 Interior Wall with a description of Fix hole in wall of guest bedroom, and a rate of $250.00.

Figure 13-1

In the Create Invoices window, enter each item you want to appear on a printed invoice. Make sure the Customer:Job and Bill To address are correct.

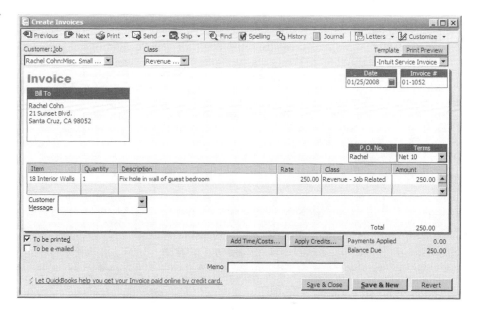

▌ Enter **Quantity** and **Rate** if necessary. Amount will be calculated automatically.

▌ From the pull-down **Terms** list, select the appropriate term. We've used Net 10 in our example.

If a term you normally use isn't on the drop-down list, you can add it. There are two types of terms — a standard term, which requires payment within a certain number of days, or a date driven term, which requires payment by a certain date. To add a standard term:

▌ From the pull-down Terms list, choose **Add New**. See our example in Figure 13-2.

▌ In the New Terms window, in **Terms**, enter a name for the new term.

▌ Click **Standard**.

▌ In **Net due in**, enter the number of days the customer has to pay the invoice.

▌ In **Discount percentage is** and **Discount if paid within**, enter a discount percentage and number of days required for the early payment discount if you offer such a discount.

▌ Click **OK** to create the new term.

Figure 13-2

Use the New Terms window to define the terms on your invoices. Use a Standard term to require payment within a number of days.

Figure 13-3

Use a Date Driven term to require payment by a certain date.

You set up a date driven term the same way, except you click Date Driven and enter the day of the month that the invoice is due. In our example, we want all invoices paid by the last day of the month, so we entered 31, as shown in Figure 13-3. You can also push back the due date if the invoice is issued late in the month, so your customers have plenty of time to pay their bills on time.

Now let's get back to the invoice. After you enter all the items you need:

▌ In the Create Invoices window, click the **Print** icon to print the invoice.

When you're satisfied the invoice is correct:

▌ Click **Save & Close** to close the invoice window. The invoice will automatically be added to Accounts Receivable.

Creating a Time-and-Materials or Cost-Plus Invoice

A time-and-materials contract means that you agree to complete a job for the documented project costs plus the labor costs (at certain rates) for your crew and yourself. A cost-plus contract means that you bill your customer for all documented costs, plus a fee (generally a percent of the documented costs). These arrangements have the least risk and the least reward. You're essentially working for wages. It's very important that you make sure all your expenses are being covered with these types of contracts. Any costs that slip through the cracks and don't get billed are a complete loss that you can't recover.

Before you actually create a time-and-materials invoice for a customer, you must enter the job costs you incurred. You must also enter timecards for the employees who worked on the job. Information on how to enter job-related costs is in Chapter 14. For information on how to enter payroll, see Chapter 15.

In this example, we'll show you how to enter a time-and-materials invoice for a job. At this point you should have already entered bills or checks to vendors and assigned the costs to the job as well as payroll, if you have employees. The example screen shots are taken from the *sample.qbw* file. If you aren't prepared to enter your own time-and-materials invoice, you can open *sample.qbw* and follow along with our example. However, as you enter data into the sample data file, the data in the file will change. Consequently, the windows you see may be somewhat different from the examples we show in this chapter.

To create a new time-and-materials invoice:

▮ Click **Home**, then click **Create Invoices**, or from the **Customers** menu, choose **Create Invoices**.

▮ From the pull-down **Customer:Job** list, select the customer. In our example, we use the Jose & Marie Sanchez: Cottage - New Construction job. If you made an estimate before preparing the invoice, you'll see a dialog box that prompts to you either select or exclude outstanding billable time and costs. Click on **Select the outstanding billable time and costs to add to this invoice**. Click **OK**.

▮ In the Choose Billable Time and Costs window, click the **Time** tab if you're charging for payroll costs and using time tracking. Put a check in the Use column to select the billable time. See Figure 13-4.

▮ In the Choose Billable Time and Costs window, click the **Items** tab. Make a checkmark in the Use column to select the billable items. In our example, shown in Figure 13-5, we checked Roof Framing, Excavation materials, Exterior Trim & Decks, and Siding.

Figure 13-4

The Time tab displays a list of all time charged to the job by employees.

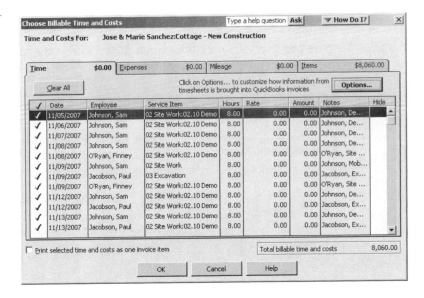

Figure 13-5

Use the Items, Expenses, and Time tabs to create the line items of a time-and-materials invoice.

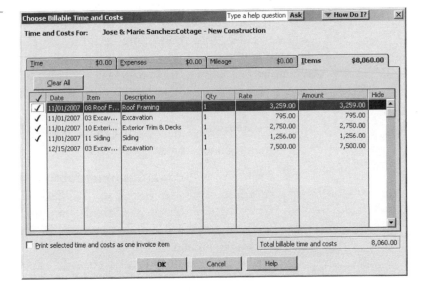

▌ If you want the selected time and costs to appear as one item on the invoice, click **Print selected time and costs as one invoice item**.

▌ Click **OK**.

▌ Back in the Create Invoices window, enter the rate you charge for the employee in the **Rate** column. Amount is calculated from the rate.

▌ Click **Print** to print the invoice.

Figure 13-6

A completed time-and-materials invoice.

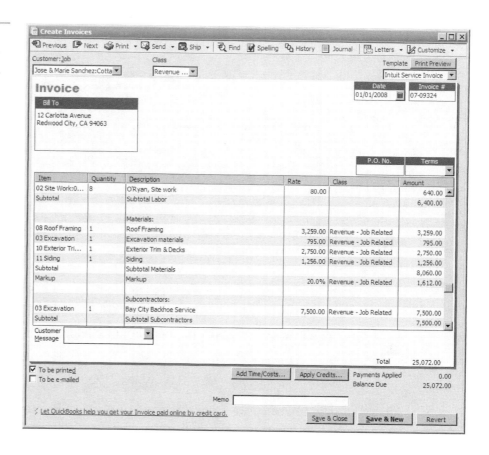

When you're satisfied that the invoice is correct:

▌ Click **Save & Close** to close the invoice window. The invoice will automatically be added to Accounts Receivable.

Your completed time and materials invoice will be similar to the example we show in Figure 13-6.

Creating a Fixed Price Contract Invoice

A fixed price (also called lump sum) contract means that, except for change orders, you agree to complete the job as specified in the plans and contract documents for an agreed amount. You assume the entire risk if the job goes over budget. But you also have the opportunity to exceed your estimated profit if you can bring the job in under the contract amount.

When you use a fixed price contract, you bill a specified amount of the contract price when you begin designated stages of the job. The specified amount is typically less than the total contract price. As an example, you might have a billing schedule like this one:

Figure 13-7
Use this item when you create an invoice for a fixed price contract.

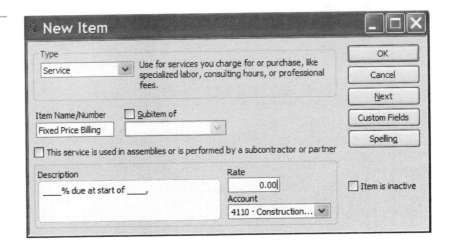

Invoice #1: 40% of contract payable at start of rough framing

Invoice #2: 20% of contract payable at start of drywall

Invoice #3: 30% of contract payable at start of final flooring

Invoice #4: 10% of contract payable at start of signing punch list

The customer or the lender usually makes an inspection to authorize a billing against the contract.

To make an invoice for a fixed price contract, you need to specify the percentage due on the invoice. And you must create an item so you can enter this percentage. Here's how to create the item:

▌ Click **Home**, then click **Items & Services**, or from the **Lists** menu, choose **Item List**. In the Item List window, pull down the **Item** menu and choose **New**.

▌ For **Type**, select **Service**.

▌ In **Item Name/Number**, enter **Fixed Price Billing**. See Figure 13-7.

▌ In **Description**, enter a generic description. This description will automatically fill in on the invoice when you use the item. In the example shown in Figure 13-7, we used "_% due at start of __." You can fill in the blanks in this description when you complete the invoice the item appears on.

▌ In **Account**, select your Construction Income account. In our example, we used 4110 Construction Income.

▌ Click **OK**.

To create a fixed price invoice you can follow our example in Figure 13-8:

▌ Click **Home**, then click **Create Invoices**, or from the **Customers** menu, choose **Create Invoices**.

Figure 13-8

A completed fixed price contract invoice.

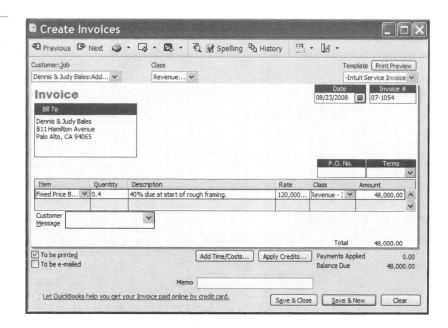

From the **Customer:Job** drop-down list, select a customer. For our example, we've used Dennis and Judy Bales: Addition. If you made an estimate before preparing the invoice, you'll see a dialog box that tells you there's an estimate for the job and asks if you want to create the invoice based on that estimate. Click **Cancel**.

From the pull-down **Class** list, select **Revenue - Job Related**.

Click in **Item** and from the pull-down **Item** list, select **Fixed Price Billing**.

In **Quantity**, enter as a decimal the percentage of the contract you're currently billing.

In **Description**, enter whatever text you need for the invoice.

In **Rate**, enter the full contract amount. Amount will be computed for you by multiplying the percentage you entered in Quantity times the full contract amount.

Click **Print**. Check the printed invoice carefully to make sure it's correct. Then print two copies so you have one for your customer file.

Click **Save & Close**.

Creating a Progress Invoice

Before you can create a progress invoice for a job, you have to do two things:

Create an estimate for the job using items. See Chapter 12 if you need information on creating estimates.

Figure 13-9

QuickBooks Pro lets you choose to invoice from an existing estimate or start the invoice from scratch.

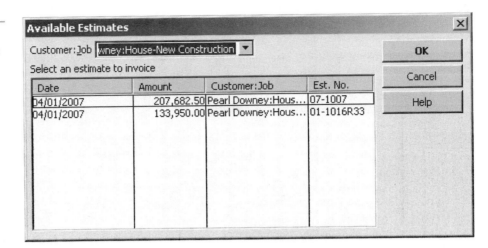

Make sure the **Progress Invoicing preference** is active.

To turn on the progress invoice preference:

▌ From the **Edit** menu, choose **Preferences** and click the **Jobs & Estimates** icon and the **Company Preferences** tab.

▌ Click **Yes** to answer the question **Do You Do Progress Invoicing?**

▌ Click **OK**.

Now, to create a progress invoice:

▌ Click **Home**, then click **Create Invoices**, or from the **Customers** menu, choose **Create Invoices**.

▌ From the drop-down **Customer:Job** list, select the customer. For our example, we've used Pearl Downey: House - New Construction.

▌ If you're following our example, you'll see a dialog box that allows you to select an estimate for the customer, as shown in Figure 13-9. At this point you'll need to select the estimate you want to create a progress invoice against.

▌ Select **OK**.

▌ This takes you to the Create Progress Invoice Based On Estimate window shown in Figure 13-10. Now you have to decide how you want to use the estimate to create the invoice. Make your selection here based on the type of contract you have with the customer. For our example, we select the last option — Create invoice for selected items or for different percentages of each item.

▌ Click **OK**.

Figure 13-10

When you use an existing estimate to create an invoice, you have three choices for how to create the invoice.

Figure 13-11

Use this window to choose which items from an estimate to include in an invoice.

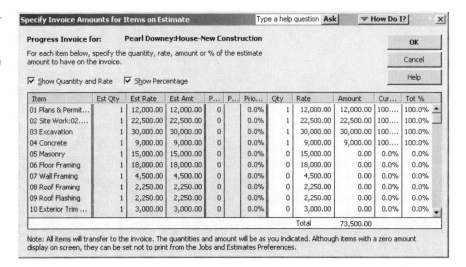

This brings up the Specify Invoice Amounts for Items on Estimate window. Here you enter the amount, percentage, quantity, and/or rate due for each item you want to appear on the invoice. To specify invoice amounts from an estimate:

▌ In the **Curr %** column for the item, enter the new percentage. Amount will fill in based on the new percentage. In our example shown in Figure 13-11, we invoiced 100% of Plans & Permits, Site Work, Excavation and Concrete.

▌ Or, in the **Amount** column for the item, enter the new amount. Curr% will fill in based on the new amount.

▌ In the **Quantity** column for the item, enter the new quantity.

▌ In the **Rate** column for the item, edit to enter the new rate.

▌ Click **OK** to get back to the Create Invoices window.

▌ Back in the Create Invoices window, click **Print** to print the invoice, shown in Figure 13-12.

Figure 13-12

This is an example of a completed progress invoice.

Check your printed invoice carefully. When you're satisfied that the invoice is correct:

▌ Click **Save & Close** to close the invoice window. The invoice will automatically be added to Accounts Receivable.

Tracking Change Orders on Estimates

With the Contractor Edition of QuickBooks, you can track change orders.

Whenever you modify items in an existing estimate, you have the option to save your changes as a change order in the description field at the bottom of the estimate form.

The change order specifies exactly what changed, the dollar amount of each change, and the net dollar change to the estimate. You may then wish to print the estimate form for the customer for final approval. (You cannot print the change order by itself.)

Figure 13-13

To create a change order in the Contractor Edition, you simply change an existing estimate.

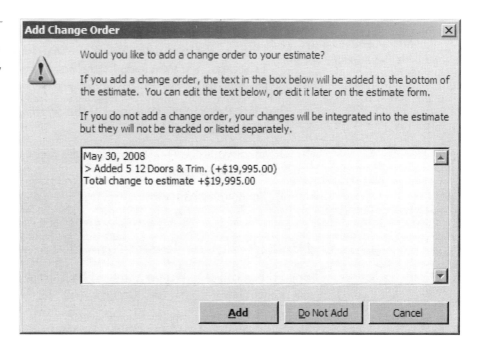

To Track Change Orders on an Estimate

1. Change your estimate as usual.

2. Save your changes.

3. QuickBooks displays the Add Change Order window, in which you can do one of the following:

▌ Click **Add** to add the displayed text to the bottom of the estimate. See Figure 13-13.

If you don't like the way the change order appears, you can edit the text in the Add Change Order window before clicking **Add**, or you can edit the change order on the estimate form later.

▌ Click **Do Not Add** if you want the estimate to be updated with your changes without explicitly listing the change order at the bottom of the form.

▌ Click **Cancel** to return to the estimate form and reconsider your changes.

Tip: If you want to save your original estimate before creating a change order, right-click in the Create Estimates window and choose Duplicate Estimate. Either save and edit the duplicate estimate, or display and edit the original estimate. Note that if you edit the duplicate estimate before you save it, a change order won't be created. Change orders are generated only on existing estimates that have previously been saved.

Reviewing and Changing an Estimate

1. Display the estimate. From the **Customer** menu, select **Create Estimates**. Hit **Previous** button until you find the estimate you want to modify.

2. Click **Print** to create a copy of the original estimate for your records. (Optional)

3. Enter your changes on the estimate form. QuickBooks automatically recalculates totals.

4. Save your changes.

How to Handle Retainage

Sometimes you'll be required to hold back 10 percent of your total progress payments (billing only for 90 percent of the contract price) as a retainer until you've completed the project satisfactorily. After the job is complete and it's agreed that you can collect the retainage, you can bill the customer for the remaining 10 percent of the project cost. Here's how to handle this type of billing:

Step 1— Setting Up an Account for Retentions Receivable

First you need to set up an asset account for retainage. If you used our example Chart of Accounts, the asset account number for retainage is 1320 - Retentions Receivable. If you don't have an asset account for retentions receivable, you'll need to follow these steps to enter a new account:

▌ Click **Home**, then click the **Chart of Accounts** icon, or from the **Lists** menu, choose **Chart of Accounts**.

▌ In the lower part of the Chart of Accounts list window, click **Account** and choose **New**.

▌ Select the account type. Click **Continue**. See Figure 13-14 for an example of a New Account window.

▌ Fill in the information for the Retentions Receivable account.

▌ Click **Save & Close**.

Step 2 — Setting Up an Item for Retentions Receivable

If you don't have an item for Retentions Receivable, follow these steps to create a new item:

▌ Click **Home**, then click the **Items & Services** icon, or from the **Lists** menu, choose **Item List**.

▌ Click **Item** in the lower part of the Items List window. From the pull-down menu choose **New**. See Figure 13-15 for an example of a New Item window.

Figure 13-14

Use the New Account window to create an account for retentions receivable.

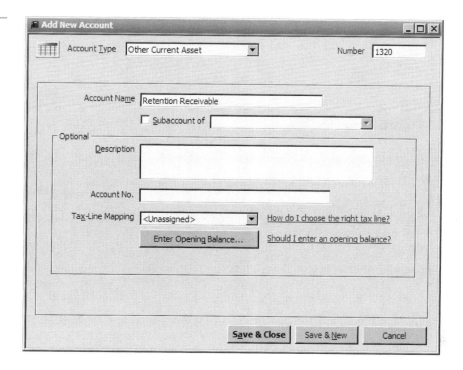

Figure 13-15

Use the New Item window to create an item for retention.

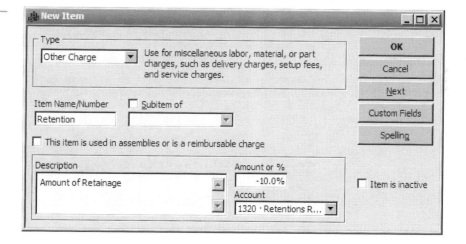

▌ In **Type**, select **Other Charge**.

▌ Fill in the fields for the Retentions Receivable item.

▌ Click **OK** to create the new item.

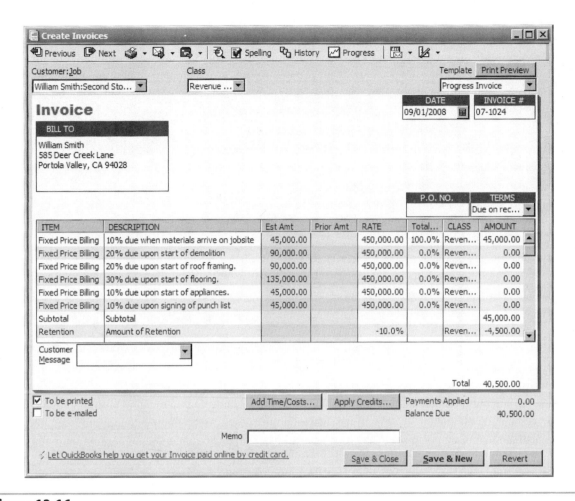

Figure 13-16
When you include a retention amount on an invoice, the balance due on the invoice is lowered by that amount.

Step 3 — Including the Retention on an Invoice

To include the retention on an invoice:

▌ Create an invoice for the job in the usual way.

▌ In the Create Invoices window, click on a blank line in the **Item** column and select the **Subtotal** item from the pull-down Item list. This will subtotal the invoice.

▌ Then select the **Retention** item from the pull-down Item list. The retention item will automatically deduct 10% from the subtotal line listed above. In Figure 13-16, the original amount of the invoice was $45,000. We entered a retainage of −$4,500. The final amount of the invoice ends up to be $45,000 − $4,500 = $40,500.

▌ Click **Save & Close**.

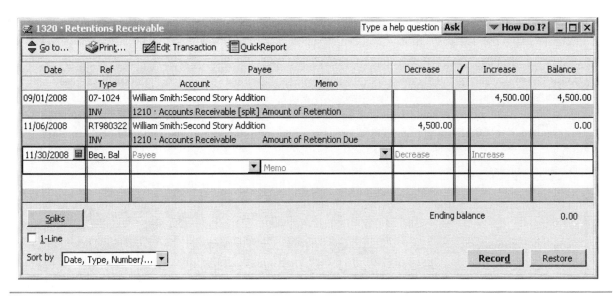

Figure 13-17
The Retentions Receivable register shows transactions in the Retentions Receivable account.

Recalling Who Owes You a Retention and How Much You're Owed

An easy way to determine the amount is to look at the register of your Retentions Receivable account. If you used our *company.qbw* data file to set up your company, you can use the memorized Retentions Receivable Report to see this information. You'll find more information on this report in Chapter 17, Reports. Otherwise, to see your retentions receivable:

 ▌ Click **Home**, then click **Chart of Accounts**, or from the **Lists** menu, choose **Chart of Accounts**.

 ▌ In the Chart of Accounts list window, select your **Retentions Receivable** account. In our example shown in Figure 13-17 it's 1320 - Retentions Receivable.

 ▌ From the bottom of the Chart of Accounts window, click **Activities** and choose **Use Register** from the pull-down menu. You should see a register that shows all the activity in your Retentions Receivable account. See Figure 13-17.

 ▌ Click the **X** in the upper right of the window to close the register.

Also, in Chapter 17, Figure 17-11, you'll learn how to print or display to screen the Retentions Receivable report. This report will summarize who owes you for retention and how much they owe.

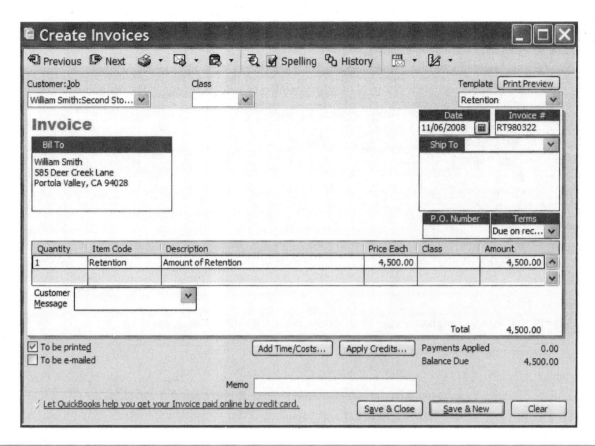

Figure 13-18
When you bill for retention, create a new invoice for just the retention.

Invoicing for Retention

When you're ready to bill for the retention, create a new invoice that contains only the retention item.

▌ Create an invoice for the job in the usual way. See Figure 13-18 for our example, using a Retention Invoice.

▌ From the Template pull-down list, select **Retention**.

▌ Click in the **Item Code** column and select **Retention** from the pull-down list.

▌ In **Amount**, enter the total amount of the retention held to date as a positive number.

▌ Click **Save & Close**.

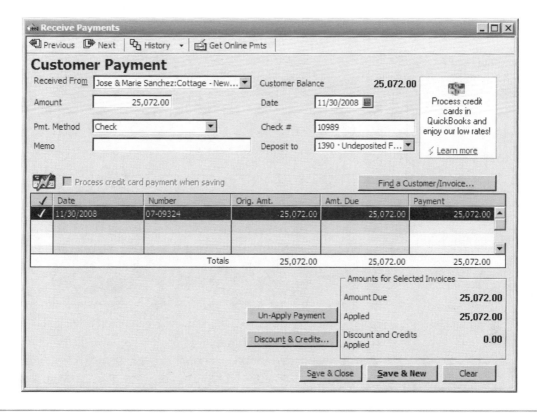

Figure 13-19
Use the Receive Payments window to record each payment you receive from a customer.

Recording a Payment You Receive

When you create and send an invoice to a customer, QuickBooks Pro creates a record in the Accounts Receivable register. When you receive a payment from the customer, you record receipt of the payment in the Receive Payments window.

To record a payment:

▌ Click **Home**, then click **Receive Payments**, or from the **Customers** menu, choose **Receive Payments**.

▌ In the Receive Payments window, select the customer from the pull-down **Received From** list. In our example shown in Figure 13-19, we show receipt of a payment from customer Jose & Marie Sanchez.

▌ Fill in the date you received the customer's payment, the amount paid, the payment method, and the check number if the customer paid by check.

Figure 13-20

Use the Payments to Deposit window to record each payment you deposit.

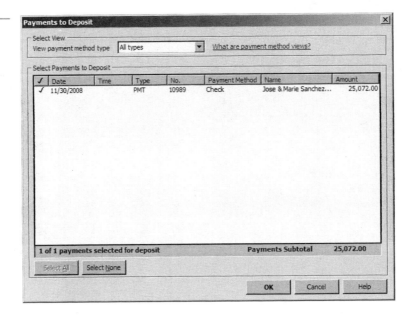

I If you want to hold the payment until you deposit the check, select your Undeposited Funds account in the **Deposit to** field. In our example, it's account 1390. We'll show you how to record the deposit in the next section. If you want to bypass the Undeposited Funds account and deposit the payment directly in a bank account, select the appropriate account in the **Deposit to** field.

I Click **Save & Close** to record the payment and close the window.

Recording a Deposit

Before you record the deposit of a customer payment, be sure you record the payment using the steps we showed in the previous section. When you record the deposit of the payment, it will be transferred from Undeposited Funds to your bank account. You can enter non-customer payments you want to deposit here.

To record a deposit:

I Click **Home**, then click **Record Deposits**, or from the **Banking** menu, choose **Make Deposits**.

> **Note:** If you don't see anything listed on the Make Deposits screen, you probably didn't select the Undeposited Funds account in the Deposit to field when you received the customer's payment (see previous section).

I In the Payments to Deposit window, click each customer check you're depositing. In our example shown in Figure 13-20, we select the payment from Jose & Marie Sanchez.

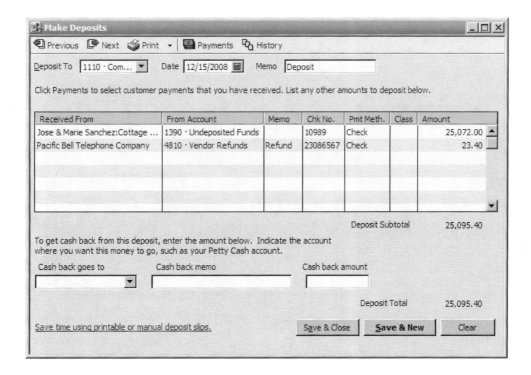

Figure 13-21
You can also use the Make Deposits window to record non-customer payments you deposit.

▌ Click **OK** to proceed to the Make Deposits window.

▌ In the Make Deposits window, add any non-customer payments that you'll be depositing. In our example shown in Figure 13-21, we deposit a refund from Pacific Bell Telephone Company for $23.40.

▌ Click **Save & Close** to record the deposit.

Recording a Job Deposit

A job deposit is money that belongs to the customer, but you hold it until you fulfill your contractual obligations. In other words, it's money in your hand, but you still need to earn it. Don't record it as income when you get it. Record it as a liability on your books even though you deposit the funds into your checking account. When the customer agrees you can apply the deposit to the job costs, or defaults on the contract, you can record the deposit as income.

To record a deposit as a liability:

▌ Click **Home**, then click **Create Sales Receipts**, or from the **Customers** menu, choose **Enter Sales Receipts**.

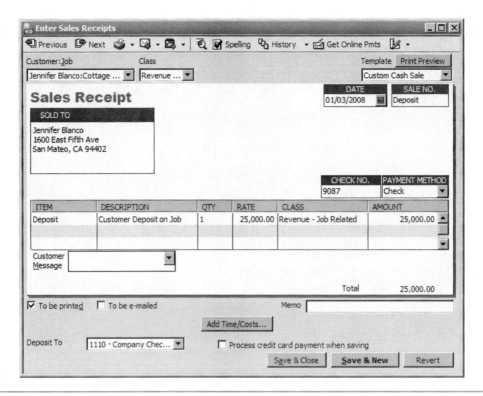

Figure 13-22
Use the Enter Sales Receipts window to record a job deposit as a liability.

▌ From the pull-down **Customer:Job** list in the Enter Sales Receipts window, select the customer. In our example shown in Figure 13-22, we enter a customer deposit for $25,000 from Jennifer Blanco.

▌ In **Class**, select **Revenue - Job Related**.

▌ In **Date**, enter the date you deposited the funds.

▌ In **Sale No.**, enter **Deposit**.

▌ Select **Payment Method** from the drop-down list.

▌ In the **Item** column, select **Deposit** from the drop-down Item list. In **Quantity**, enter **1**. Enter **Rate**. Amount will be calculated.

▌ To print a message on the receipt, select a message from the **Customer Message** drop-down list. To add a new message, click **Add New** and create your own personalized message.

▌ Click the **Print** icon to create the receipt for your customer.

▌ At the very bottom of the Enter Sales Receipts window, click **Deposit To**. From the pull-down Account list, select the **Checking** or **Savings** account you want to put the funds in. Make sure you deposit

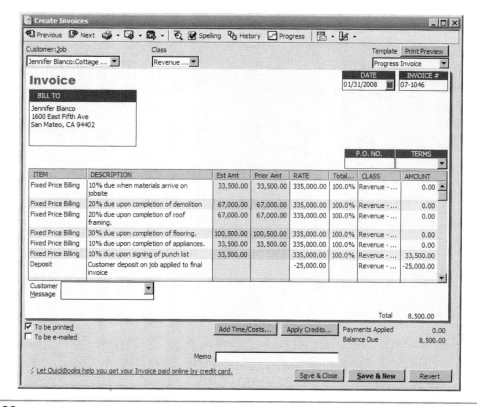

Figure 13-23
This Create Invoices window shows a deposit deducted from a final billing.

this check by itself, not as part of a group deposit. When you reconcile the Checking or Savings account in QuickBooks Pro, you want this deposit to be listed alone.

▌ Click **Save & Close** to record the deposit.

To apply the deposit to an invoice:

▌ Follow the usual steps for creating an invoice for the customer.

▌ At the end of the invoice items, enter a deposit item using a negative rate. In our example shown in Figure 13-23, we subtracted $25,000 from the Jennifer Blanco:Cottage Progress invoice.

▌ Click **Save & Close** to create the invoice.

Now you know how to process invoices to customers in QuickBooks Pro.

Chapter 14

Payables

We've based the information in this chapter on using QuickBooks Pro on the accrual method of accounting we discussed in Chapter 13. The advantages of the accrual method of accounting, as it relates to Accounts Payable, are:

▌ your job cost reports will include vendor bills that you haven't paid

▌ your Accounts Payable Report will help you track the vendors you owe money to

▌ your profitability reports will be more accurate

If you don't use an accrual method of accounting, you'll find the information you need on cash accounting in Chapter 16.

Here's what we're going to discuss in this chapter:

1. Entering purchase orders

2. Using purchase orders to track multiple draws and committed costs

3. Entering bills for job-related and overhead expenses without purchase orders

4. Selecting bills for payment

5. Printing checks

6. Payables and vendor workers' comp reports

To get job and non-job expenses into your accounting system accurately and completely, you'll have to enter who you're paying and what you're paying for. Be sure to enter a memo or note about each expense and put a record of the transaction in the proper account. Then your QuickBooks Pro reports will accurately reflect all your expenses.

You should already have set up your Chart of Accounts and other lists in QuickBooks Pro before going through this chapter. You should also open the *sample.qbw* file and use it to follow the examples in this chapter. Then you can try out the steps you need to go through without affecting any actual company information.

Figure 14-1

Use purchase orders to track items that you've ordered and check against what you receive.

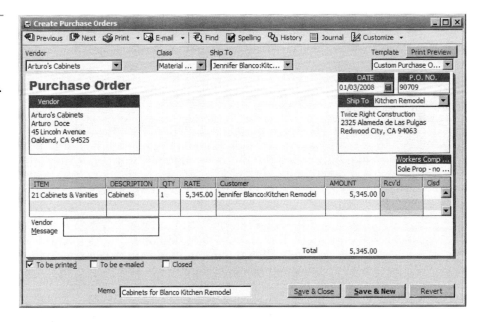

Creating and Using Purchase Orders

Purchase orders in QuickBooks Pro are a record of what you've ordered. They let you compare what you order with what you actually get. For example, suppose Twice Right Construction ordered cabinets from Arturo's Cabinets and created the purchase order shown in Figure 14-1. QuickBooks Pro will track the order date and the amount quoted for the cabinets. Then Twice Right doesn't have to rely on memory for this information.

So let's look at how to create a purchase order:

▌ Click **Home**, then click the **Purchase Orders** icon, or from the **Vendors** menu, choose **Create Purchase Orders**.

If you don't see the command Create Purchase Orders in the Vendors menu, you need to turn on purchase order tracking. To do this:

▌ From the **Edit** menu, choose **Preferences**.

▌ Click **Items & Inventory**.

▌ Click the **Company Preferences** tab.

▌ Select **Inventory and purchase orders are active**.

▌ Click **OK**.

Now, back to our purchase order.

▌ In the top part of the form, enter the Vendor, the Ship To address (job site), the Class (materials), Date, PO number, and vendor's address.

Figure 14-2

Use a purchase order to enter the parts of a multiple-payment draw schedule.

▌ Click in the **Item** column and select the items you're ordering from the pull-down list. Change the Description and Amount fields as necessary.

▌ Click in the **Customer** column and select the customer you're ordering materials for from the pull-down Customer list.

▌ In **Memo** enter a brief description about this PO. This is important because the memo will appear when you move the PO into a Vendor bill.

▌ Click **Save & Close** or **Save & New** to enter another PO.

Using Purchase Orders to Track Multiple Draws and Committed Costs (Buyouts)

In QuickBooks Pro, you can assign each line item of a purchase order to a job. This feature will let you use the purchase order to track multiple payment draws and committed costs to subcontractors. Let's look at multiple draws first.

Multiple Draw Schedules

Suppose you want to track an agreement like the one we made with Ace Framing Contractors for a three-part draw. As soon as you make this type of agreement or contract, you should enter the draw schedule as a purchase order. In Figure 14-2 you'll see an example showing the purchase order where Twice Right Construction entered a three-part draw to Ace Framing for floor (Draw #1), wall (Draw #2), and roof (Draw #3) framing.

Figure 14-3

Use the Open Purchase Orders window to see a list of the open POs for a vendor.

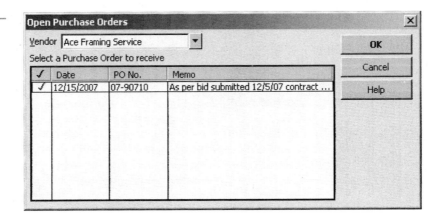

When Twice Right received Ace Framing's first bill for Draw #1 — Floor Framing, they used that same purchase order to record the payment of the first draw. Here are the steps:

▌ Click **Home**, then click **Receive Inventory**, **Receive Inventory with Bill**, or from the **Vendors** menu, choose **Receive Items and Enter Bill**.

▌ In the Enter Bills window, from the pull-down **Vendor** list, select the appropriate vendor. In our example it's Ace Framing.

▌ If you're following our example, you'll see the Open PO's Exist window, where you'll be warned that open purchase orders exist for the vendor. You're asked if you want to receive against one of them. Click **Yes** here to tell QuickBooks Pro that you want to use an existing PO.

▌ In the Open Purchase Orders window, the vendor you selected appears along with a list of the purchase orders open for the vendor. Click in the ✓ column to select the appropriate open PO. In our example shown in Figure 14-3, we checked the PO dated 12/15/2007.

▌ Click **OK**.

Back in the Enter Bills window, as shown in Figure 14-4:

▌ In **Date**, enter the date on the vendor's bill.

▌ In **Ref. No.**, enter the vendor's invoice number.

▌ In **Amount Due,** enter the amount you plan to pay the vendor (or the draw). In our example, Twice Right will pay $3000, which is 100 percent of Draw #1 - Floor Framing.

▌ In **Qty,** enter 1 for any draw that you want to pay now. If you only want to pay half of the draw, enter 0.5 in Qty. If you don't want to pay any of the draw, enter 0.

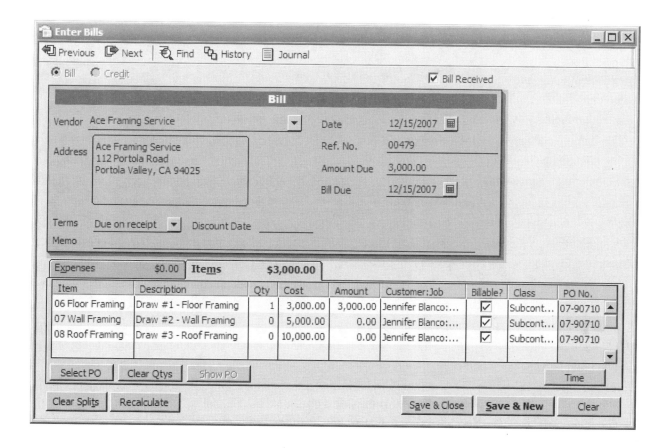

Figure 14-4
In this Enter Bills window, Twice Right will pay the first draw to Ace Framing.

▌ The **Amount** column will calculate automatically. In our example, it's $3,000 for Draw #1. The amount shown on the Items tab should be the same as the amount you're paying. The amounts for the other draws you're not paying will be 0.00.

▌ Click **Save & Close**.

Committed Costs

You can also use a purchase order and the same method to track what you've committed to pay a subcontractor. For example, suppose you have a subcontractor who bid a project out at a certain price and you want to track the bid price to compare it to his final bill. To do this, you could set up a purchase order at the beginning of the project and compare it to the final bill

Figure 14-5

Use a PO to track a bid price from a subcontractor.

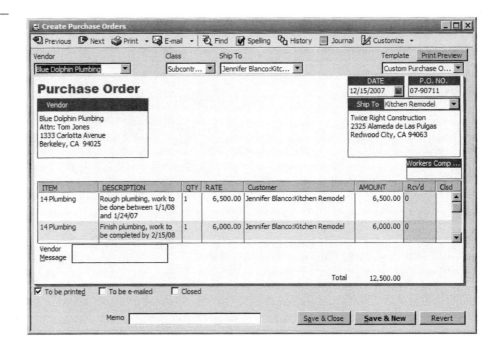

you receive from the subcontractor. In our example in Figure 14-5, Twice Right used a PO to track the bid price (and/or contract price) from Blue Dolphin Plumbing for rough and finish plumbing. When Blue Dolphin Plumbing submits their bill, Twice Right can compare the original bid price to the original purchase order.

Tracking Unfilled Purchase Orders

QuickBooks Pro has a built-in report to track purchase orders that haven't been filled. You can run this report to list all subcontractors who haven't completed or submitted bills on a project. To generate this report:

▌ Click the **Report Center**, or from the **Reports** menu, choose **Purchases** and then **Open Purchase Orders by Job**.

▌ In the Modify Report: Open Purchase Orders by Job window, click **OK**.

Important note! You may have a problem with this report if you make a partial payment to a subcontractor. The amount remaining after the partial payment will clear from the report unless you enter a number less than 1 in the Qty column in the Enter Bills window. To avoid this reporting problem, always enter a number *less than 1* when you release part of a PO. For example if you pay $1,500 of a $3,000 bid, enter 0.5 in Qty.

Figure 14-6

Use the Items tab on the Enter Bills window to record job-related expenses.

Entering Bills Without Purchase Orders

Entering Bills for Job-Related Expenses

Suppose you don't have the need (or time) to enter a purchase order before you get a vendor bill. In this case, you'll want to skip entering the purchase order and enter the bill directly. In Figure 14-6 we show an example of a bill Twice Right got from Eric Savage Electrical Company for $2,300 that they hadn't entered a purchase order for. To record the $2,300 they owe and get the fee into their job cost reports, they'll enter a bill in QuickBooks Pro. It's important to enter job-related costs on the Items tab when you enter a bill.

To enter a bill for a job-related expense:

▌ Click **Home,** then click **Enter Bills**, or from the **Vendors** menu, choose **Enter Bills**.

▌ Fill in the **Bill** form in the top half of the window.

▌ In **Memo**, enter a memo that tells what the bill is for. This memo will appear on billing reports such as A/P Aging Summary and Vendor Balance Detail. If you pay the bill by check and you *didn't* enter an account number when you set up the vendor's account, QuickBooks Pro puts the memo in the Memo field on the check. If you *did* enter an account number when you set up the vendor's account, QuickBooks Pro puts the account number in the Memo field of the check.

▌ Click the **Items** tab. Be sure that you enter job-related expenses on the Items tab of the Enter Bills window.

Figure 14-7

Use the Expenses tab on the Enter Bills window to record overhead expenses. Make sure you enter the correct account.

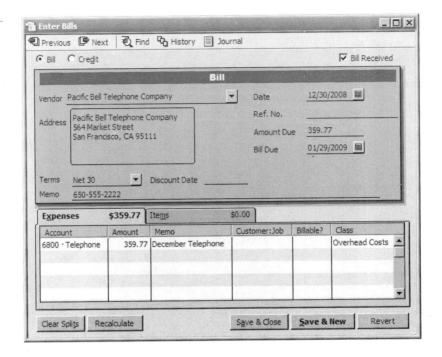

- Click in the **Item** column and select the items for the bill from the pull-down list.

- Click in the **Customer:Job** column and select the customer from the pull-down list.

- Click in the **Class** column and select the class from the pull-down list.

- Edit **Description** and **Amount** if necessary.

- Click **Save & Close** to record the bill.

Entering Bills for Overhead Expenses

Non-job related expenses are overhead expenses for things such as telephone, electricity, office supplies, etc. You should enter a bill for any expense when you receive the bill, even if you charged it to a credit card or you have an account with the company. Then you'll have the costs in your books as soon as it's been charged.

To enter a bill for an expense that's not a job-related expense:

- Click **Home,** then click **Enter Bills**, or from the **Vendors** menu, choose **Enter Bills**. In our example shown in Figure 14-7, we show a bill from an electrician.

- Fill in the **Bill** form in the top half of the window.

▌ In **Memo** (not the Memo column on the Expenses tab), enter a memo that tells what the bill is for. This memo will appear on billing reports such as the A/P Aging Summary. If you pay the bill by check and you *didn't* enter an account number when you set up the vendor's account, QuickBooks Pro puts the memo in the Memo field on the check. If you *did* enter an account number when you set up the vendor's account, QuickBooks Pro puts the account number in the Memo field of the check.

▌ Click the **Expenses** tab. It's important that you use the Expenses tab of the Enter Bills window for non-job expenses. You have to make sure each bill you enter is associated with the proper account. That's the way to get useful and accurate reports on your expenses.

▌ Click in the **Account** column and select the appropriate expense account or accounts to assign the bill from the pull-down list. If you're paying a bill with multiple items and you need to split the amount over more than one expense account, change the amount that's automatically entered in the first line to match the line item in the bill. Then tab to the second line and make the entry for the next line item. Continue this way for each line item.

▌ In the **Memo** column, enter a description or account number (you won't have a Customer:Job for non-job related expenses).

▌ Click **Save & Close** to record the bill.

Selecting Bills for Payment

The Pay Bills feature in QuickBooks Pro lets you select the bills you want to pay. This is usually called a check run. In the construction industry, check runs usually happen every Friday or on the last day of the current month and the 15th day of the following month. In this section, we'll explain how to select the bills you want to pay. In the next section, we'll explain how to print the checks, if you have check stock.

To select bills for payment:

▌ Click **Home,** then click **Pay Bills**, or from the **Vendors** menu, choose **Pay Bills**. QuickBooks Pro will list all unpaid bills or all bills due for the date range you enter.

▌ At the top of the Pay Bills window, next to **Show bills,** select **Due on or before** to enter a date to cover all the bills that you want to pay. Or select **Show all bills**.

▌ Click in the ✓ column to select each bill you want to pay. As you select the bills, compare the total in the **Amt To Pay** column against the total shown for **Ending Balance** in the Payment Account section of the Pay Bills window. This tells you if you have enough money in your checking account to cover all the bills you've selected to pay.

Figure 14-8

Select the bills you want to pay in the Pay Bills window.

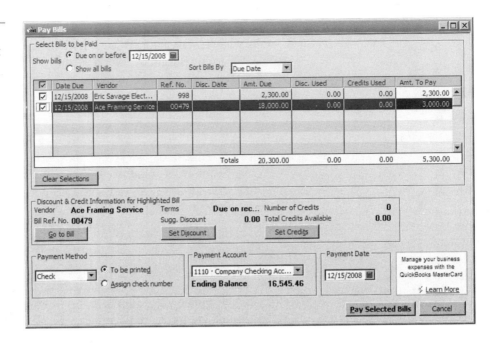

- To pay only part of a particular bill, change the amount shown in the **Amt To Pay** column. In our example shown in Figure 14-8, Twice Right paid $3,000 of the $18,000 owed to Ace Framing Service.

- In the Payment Method section of the Pay Bills window select **To be printed**. From the pull-down Payment Method list, select **Check**.

- To record the payments, click **Pay Selected Bills**.

- The Payment Summary window will appear. If you're ready to print checks, click the **Print Checks** button. Otherwise, click **Done**.

Printing Checks

To print your checks:

- From the **File** menu, choose **Print Forms** and then **Checks**. or click the **Print Checks** icon.

- Put a check mark in front of each check you want to print. See Figure 14-9.

- Click **OK** to print the checks.

Be sure you have blank check stock in your printer, and that the number in the First Check Number field matches the first number on your checks in the printer.

Figure 14-9

Select the checks you want to print in the Select Checks to Print window.

Vendor Workers' Comp Reports

Staying on top of the workers' comp expiration dates for your subcontractors is a constant headache. As a matter of policy, you shouldn't even issue a purchase order, let alone pay a bill, for a subcontractor who hasn't supplied you with a current certificate of insurance for workers' comp. A workers' comp audit can cost you dearly if you can't show that your subcontractors had coverage at the time they provided their services. You can easily create a custom field in your vendors list to hold the date each subcontractor's workers' comp expires. You may find it helpful to get a report of your vendors and their workers' comp expiration dates. The *sample.qbw* file on the CD with this book has the custom field Workers' Comp Expires created for vendors. You'll find more information on the Vendor Workers' Comp Expirations report in Chapter 17, Reports.

Here's how to create a custom field:

- Click the **Vendor Center** icon on the menu bar, or from the **Vendors** menu, choose **Vendor Center**.

- In the **Vendor Center** window, right click any vendor and choose **Edit Vendor**.

- Click the **Additional Info** tab. The example in Figure 14-10 shows the Workers' Comp Expires custom field.

- In the lower right of the Custom Fields area, click **Define Fields**.

- In the Define Fields window, under the Label column, enter the field names you want to create. See Figure 14-11.

- In the columns to the right, check the boxes to select who you're using the custom field for.

- Click **OK** in the Define Fields window.

If you're using the QuickBooks Premier Contractor edition, you don't need to create this custom field. You can track insurance expiration dates on the Additional Info tab. See Figure 8-5 in Chapter 8, Vendors & Subcontractors.

Figure 14-10

Use the Additional Info tab on the Edit Vendor window to create custom fields.

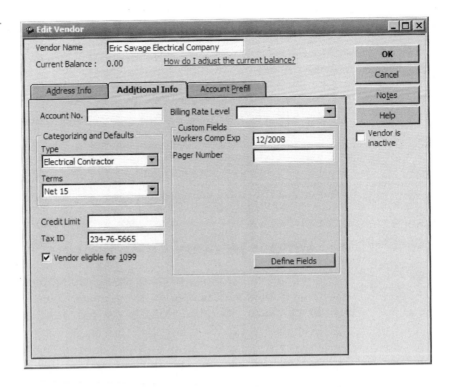

Figure 14-11

Use the Define Fields window to name a custom field and select who the field will be used for.

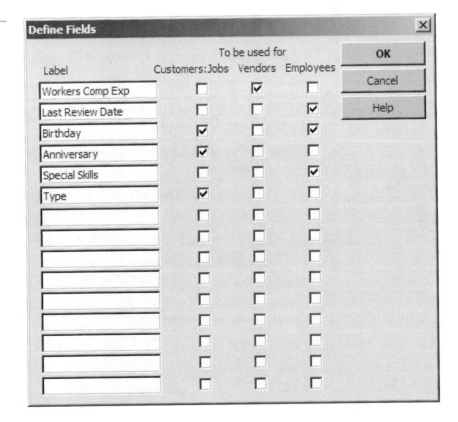

Figure 14-12

The Templates list shows the existing templates.

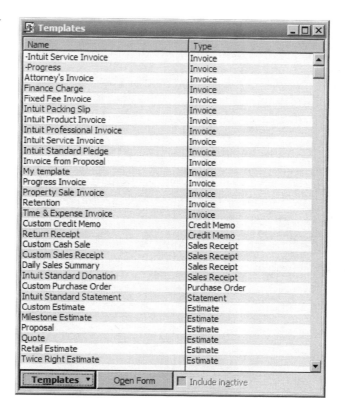

Adding a Custom Field to Purchase Order Forms

To add a custom field to your purchase order forms:

▌ From the **Lists** menu, choose **Templates**. In Figure 14-12, we show the Templates list.

▌ Double click **Custom Purchase Order** in the **Name** column.

▌ Click **Additional Customization** to get a window like the one shown in Figure 14-13.

▌ Select the check box next to **Workers Comp**. You can also specify if you want the custom field printed on the purchase order, if you want it to appear only on the screen, or if you want both. Usually you'll choose to see the custom field only on the screen.

▌ Click **OK** to add the field to the purchase order form.

Preparing a Vendor Report on a Custom Field

Here's how to make a vendor report on a custom field:

▌ Click the **Report Center**, or from the **Reports** menu, choose **Vendors & Payables**, then **Vendor Contact List**.

Figure 14-13
Use the Fields tab window to select which field you want to see in the Purchase Order window and which fields you want printed on your purchase orders.

▌ On the **Display** tab of the Modify Report: Vendor Contact List window, under the **Columns** box, scroll through the list to select only **(left margin)**, **Vendor**, **Phone**, **Contact**, and **Workers Comp Exp**. See Figure 14-14.

▌ Click **OK** to see a report like the Vendor Contact List we show in Figure 14-15.

Now that you've finished entering bills and writing checks, you can proceed to payroll, Chapter 15. Even if you don't have any employees, you should read through this chapter because we've included a special section on how to enter time for sole proprietors and partnerships.

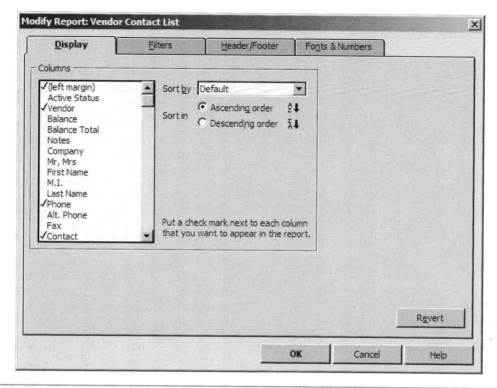

Figure 14-14

Use the Display tab of the Modify Report: Vendor Contact List window to select the columns you want to appear in a Vendor Contact list.

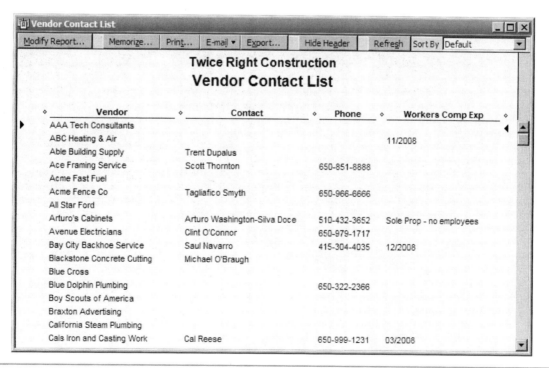

Figure 14-15

This Vendor Contact List report shows the custom field Workers' Comp Expires.

Payroll

QuickBooks Pro's ability to handle payroll is one of its best features. It has tremendous flexibility, which is also why setting it up can be somewhat frustrating.

In this chapter we'll show you how to:

▌ enter a timesheet

▌ process payroll

▌ print payroll checks

And, if your company is a sole proprietorship or partnership, we'll show you how to enter and process owner's time to jobs so it shows on your job cost reports. But before we begin, make sure you have all your payroll items set up. You'll find information on payroll items back in Chapter 5.

Entering a Timesheet

Before creating a payroll check for any employee, you need to complete a pay period timesheet to assign the hours the employee worked to a customer, job, service item (job phase), and payroll item. You get this information from the weekly timesheet filled out by the employee, foreman, or supervisor in charge of the project. Some states require that employees complete their own timecards.

Before you process payroll the first time, we suggest you open the *sample.qbw* file that comes with this book. We'll walk you through entering a weekly timesheet and processing payroll. The employee we'll be using first is Joe Bliss. In our sample company world, he's filled out and turned in a timecard for the week ending December 27, 2009. To begin, we're going to enter the information on his timecard into QuickBooks Pro.

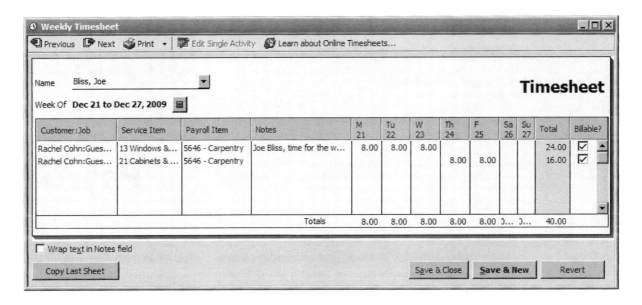

Figure 15-1
The QuickBooks Pro timesheet assigns the hours worked to a Customer:Job, Service Item, and Payroll Item.

To enter a timesheet:

▍ Click **Home**, and **Enter Time**, then click the drop arrow and select **Use Weekly Timesheet**, or from the **Employees** menu, choose **Enter Time**, and then **Use Weekly Timesheet**.

▍ From the **Name** drop-down list, select the employee name. Figure 15-1 shows our Weekly Timesheet example for employee Joe Bliss.

You'll see the current week in the upper left-hand corner of the window. You have to make sure you're entering the time for the correct week. To move to different weeks:

▍ Click **Previous** or **Next**. To quickly move to a particular date, click the calendar icon at the end of the "Week of" field. In our example, it's the week of December 21, 2009.

Let's look at the Weekly Timesheet in Figure 15-1 in detail.

The Customer:Job column represents the customer and job. As you see, Joe Bliss worked on the Rachel Cohn:Guest Cottage.

The Service Item column represents the job phases the employee worked on. For more information on job phases, see Chapter 4, Items. In our example, Joe worked on different job phases during the week. On Monday through Wednesday he worked on Rachel Cohn:Guest Cottage installing windows and trim. Then on Thursday and Friday Joe installed cabinets.

Figure 15-2

Warning window for a payroll item that hasn't been set up.

The contents of the Payroll Item column depend on whether or not you're using payroll items to track workers' compensation. In our example, we use workers' compensation codes as Payroll Items in QuickBooks Pro. This lets us print a report later to help us fill out our workers' compensation report.

Entering a Payroll Item Not on an Employee's Payroll Info Tab

If you enter a payroll item that isn't on the employee's Payroll Info tab, you'll get a warning like the one shown in Figure 15-2. The Payroll Info tab is a part of an employee's record. To fix this problem:

▌ Click the **Employee Center** on the menu bar, or from the **Employees** menu, choose **Employee Center**.

▌ In the Employee Center window, double-click the employee you're entering the timecard for.

▌ From the **Change Tabs** dropdown list, choose **Payroll and Compensation Info**.

▌ In the Earnings section of the Edit Employee window, click in the **Item Name** column and select the item(s) you need from the pull-down Payroll Item list. Our example in Figure 15-3 shows the Payroll Info tab for Joe Bliss.

▌ Click **OK**.

Returning to the Weekly Timesheet, the Notes column is extremely important if you'll be billing a customer for an employee's labor on a time-and-materials basis. If you use QuickBooks Pro to generate an invoice for the customer, make sure you fill in Notes with the exact text you want the customer to see on the invoice. The text you enter here will appear as a description on the invoice.

To the right of the Total column is the Billable column. Put a checkmark in the box for any time that is billable on a time-and-materials job.

Figure 15-3

Payroll items for an employee are listed in the Earnings box of the employee's data record.

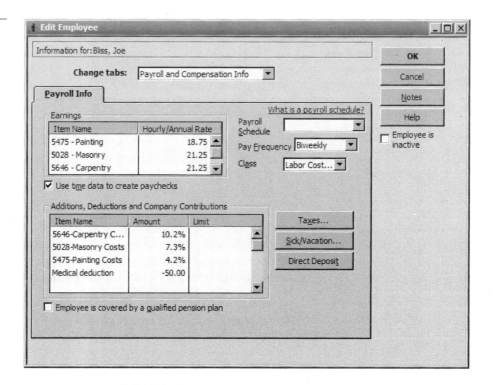

Now back to what to enter on the timesheet:

▌ Enter on a different line the hours worked on each job, phase, and (if needed) workers' comp code.

▌ Enter the time for the week.

▌ Click the billable column boxes as necessary for the jobs.

▌ Click **Save & Close** to record the timesheet.

Any report generated before processing payroll will show only the hours entered, not the dollar amounts for those hours. After processing payroll, the dollar amounts associated with the hours will appear on reports.

Processing Employee Payroll

Before you can process payroll for an employee, you must enter a timesheet for the employee. And before you can process payroll for the first time you need to go through the steps outlined below to set up the payroll module. To process payroll the first time for employees you've entered timesheets for:

▌ From the **Employees** menu, choose **Payroll Service Options**, then **Learn About Payroll Options**.

QuickBooks will launch a web browser and display payroll options.

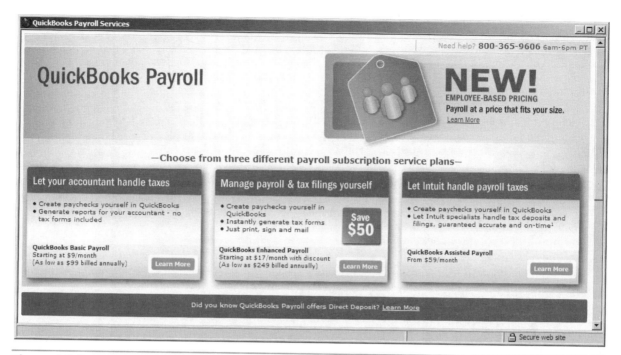

Figure 15-4
The QuickBooks Payroll Services window displays all your payroll options.

At the QuickBooks Payroll Services window (see Figure 15-4), you have the following options:

1. *Basic Payroll* — Calculates deductions, earnings, and payroll taxes for up to 3 employees.

2. *Enhanced Payroll* — All the features of Basic Payroll, plus automatically fills in quarterly Federal and State tax forms.

3. *Assisted Payroll* — All the features of Enhanced Payroll, plus payroll tax deposit and filing, W-2 printing and mailing, and a "no penalties" guarantee.

4. *Manual Calculations* - Select this option if you prefer to manually calculate payroll, or if you use a payroll service and just want to use QuickBooks payroll to get labor costs included in job costs. To enable manual payroll:

   ▌ From the **Help** menu, select **QuickBooks Help**.

   ▌ Click on the **Search** tab. Type in **process payroll manually** and click on the arrow at the end of the search box.

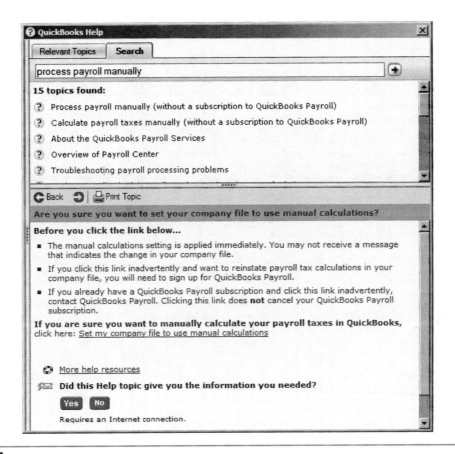

Figure 15-5
If you have a payroll service and want to get your labor cost to jobs, set up the QuickBooks Pro payroll module to manually calculate payroll.

▌ This will bring up a list of topics. Click on **Process payroll manually (without a subscription to QuickBooks payroll)**.

▌ In the lower section of the **Help** window, locate step #2 and click on **manual payroll calculations**.

▌ Below the bulleted list, you will see the sentence **If you are sure you want to manually calculate your payroll taxes in QuickBooks, click here**. Click on **Set my company file to use manual calculations**.

If you select this option, you need to manually calculate the payroll taxes for each employee every time you run a payroll check.

▌ Click **Home**, then click the **Pay Employees** icon, or from the **Employees** menu, select **Pay Employees**.

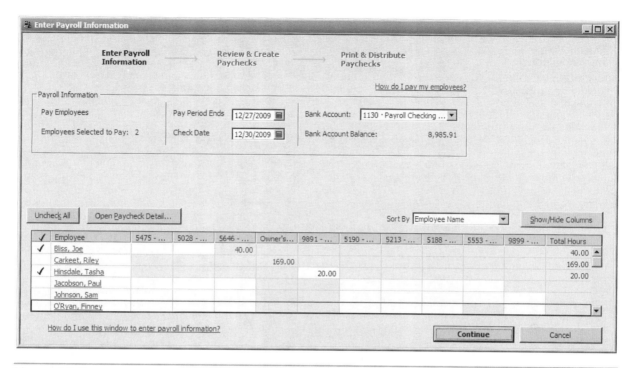

Figure 15-6
Put a check mark next to each employee to be paid.

▌ In the Enter Payroll Information window, select the correct Bank Account you'll use for the checks. In our example shown in Figure 15-6, we used our 1130 Payroll Checking Account.

▌ Enter **Check Date** and **Pay Period Ends** date.

If hours don't show up for an employee even though you know you've entered time for that employee, you probably didn't enter the pay period ending date correctly. To make sure all time you entered on a previous timesheet shows up, you must enter a date after the last date that shows up on the previous timesheet you entered. For example, in Figure 15-1, the last date on the timesheet next to Week Of is 12/27/2009. To pick up all the time on that timesheet, the date you enter into the Pay Period Ends box must be 12/27/2009. Also, the date must not conflict with the next timesheet you enter.

▌ Click in the leftmost column next to the employee name for each employee you're paying.

▌ Next you'll review each paycheck for accuracy.

Figure 15-7

Use the Preview Paycheck window to make sure you enter all costs for an employee correctly.

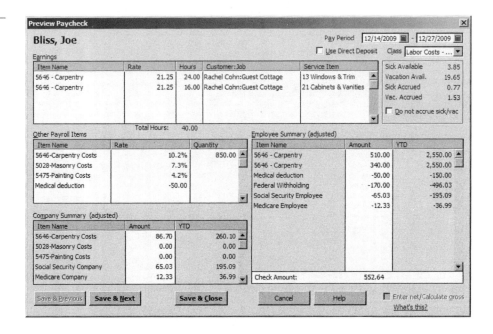

Using the Preview Paycheck Window

The Preview Paycheck window allows you to review each person's paycheck for accuracy. Under the Employee column of the Enter Payroll Information window, click on each employee name to get to the Preview Paycheck window.

The Preview Paycheck window displays the information that was entered on the timecard.

The hours for each payroll Item, Customer:Job, and Service Item you entered will show in the Earnings box. The information in the Earnings box comes directly from the timecard information you input onto the Weekly Timesheet. Our example shown in Figure 15-7 is the Preview Paycheck window we get using the timesheet shown in Figure 15-1.

You must make sure here that the information in the Earnings box is correct. If any information isn't correct, you should click Cancel to close the Preview Paycheck window right away. Then return to the employee's timesheet you entered and make the necessary changes. Also, if you make changes in the Earnings box in the Preview Paycheck window, the timecard won't be automatically changed. You must make any corrections directly on the timesheet.

Entering Payroll Additions and Deductions

You use the Other Payroll Items box to track additions to an employee's pay (bonus), deductions from an employee's pay (medical or dental), and any expenses that the company has because of the employee's paycheck (workers' compensation expense). In our example, we deducted $50 from Joe's paycheck for medical insurance. To do this, in the Preview Paycheck window:

▌ Click in the **Item Name** column under the **Other Payroll Items** box and select the addition or deduction from the pull-down Item Name list.

▌ In the **Rate** column, enter the amount of the addition or deduction. Check to make sure you've entered the information correctly. To enter a deduction, enter a negative number in the **Rate** column. In our example in Figure 15-7, we deducted $50 for medical insurance from Joe Bliss's check.

▌ Be sure to check your entries in the Employee Summary box.

If you want to send the cost of workers' compensation to the Customer:Jobs listed in the Earnings Box, you'll need to enter the workers' compensation codes (company contributions payroll items) under the Other Payroll Items box. If you haven't set up these payroll items, you'll find information on how to do this back in Chapter 5.

You also use the Preview Paycheck window to add the cost of workers' compensation to jobs. To do this:

▌ In the **Item Name** column, under the **Other Payroll Items** box, select the workers' compensation code that matches the first line listed in the Employee Summary box. In our example, Joe has 5646-Carpentry Costs listed as the first line in the Employee Summary box. We select its matching company contribution payroll item in the Item Name column, 5646-Carpentry Costs. The rate will appear automatically, but you'll need to fill in **Quantity**. In our example, Quantity is the total of all wages 5646-Carpentry Costs in the **Amount** column of the Employee Summary box. 5646-Carpentry Costs appears two times. So Quantity is $510 + $340 which is $850.

▌ Continue entering a matching workers' compensation code for each payroll item listed in the Employee Summary box that's subject to workers' comp.

If you follow this procedure, the cost of the workers' compensation insurance will be spread to each of the jobs listed in the Earnings box.

Figure 15-8
Print Paychecks window.

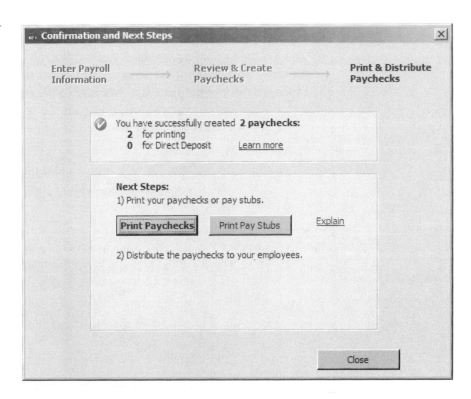

Check the employee's information in the Preview Paycheck window again. If everything is correct:

▌ Click **Save & Next**. Preview each paycheck.

Once all paychecks have been reviewed:

▌ Click **Continue** in the **Enter Payroll Information** window.

▌ Give the paychecks a final review in the **Review and Create Paychecks** window.

▌ When everything looks correct, click **Create Paychecks**.

▌ A payroll confirmation window is displayed. See Figure 15-8.

▌ If you're ready to print the paychecks, click **Print Paychecks**. If not, click **Close**. Follow the steps below to print the checks when you are ready.

Printing Employee Checks

To print the employee checks:

▌ From the **File** menu, choose **Print Forms** and then **Paychecks**.

▌ Select the paychecks to print, as shown in our example in Figure 15-9.

▌ Click **OK** to print the checks.

Figure 15-9
Put a check mark next to each check you're ready to print.

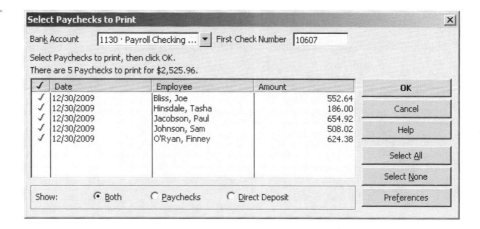

Allocating Sole Proprietor or Partner's Time to a Job

One of the most common questions we hear is "How do I, as an owner, allocate my time to a job?" All you have to do is enter your time on a timecard and then process payroll. While you process payroll, you deduct the dollars back out by creating a zero check for yourself. Entering the timecard and processing payroll will add the hours, and the dollars associated with those hours, to your job cost reports. But be sure you've set up payroll items for owner's wages and owner's labor deduction. See Chapter 5 if you need help setting up a payroll item.

Entering a Timesheet for an Owner

For each pay period, enter the owner's time on a timesheet just as you would a regular employee. Let's go through an example for Riley Carkeet, the sole proprietor of Twice Right Construction. To enter a timesheet for an owner or partner:

▌ Click **Home**, and **Enter Time**, then click the dropdown arrow and select **Use Weekly Timesheet**, or from the **Employees** menu, choose **Enter Time**, and then **Use Weekly Timesheet**.

▌ From the **Name** drop-down list, select the owner. Figure 15-10 shows our Weekly Timesheet example for Riley Carkeet.

▌ Select the appropriate week. In our example, it's December 21 - December 27, 2009.

▌ Enter a different line for the hours worked on each job and phase.

▌ In the **Payroll Item** column of each line on the timesheet, make sure you select the payroll item **Owner's Labor**. You must always use this item when you enter owner's time. The Owner's Labor item is set up to post to an equity account (account 3999 Owner's Time to Jobs in

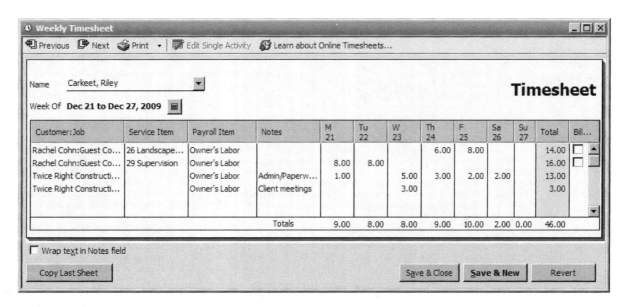

Figure 15-10
When entering a timesheet for an owner, the payroll item must be Owner's Labor.

our *sample.qbw* file). If you select a payroll item that doesn't post to the appropriate equity account, the owner's labor costs won't appear properly in your job cost reports.

▌ Click the invoice icons as necessary for the jobs.

▌ Click **Save & Close** to record the owner's time.

Processing an Owner's Time

To process payroll allocating owner's time that you have entered a timesheet for:

▌ Click **Home,** then click the **Pay Employees** icon, or from the **Employees** menu, choose **Pay Employees**.

▌ From the pull-down Bank Account list, select your **Adjustment Register** account. Since you're creating a zero check, you don't want it showing up in the regular bank account. Use the Adjustment Register to hold transactions with zero amounts. In our example shown in Figure 15-11, we use 1111 Adjustment Register account.

▌ Enter the **Check Date** and **Pay Period Ends**.

▌ Click in the checkmark column next to the owner's name.

▌ Under the Employee column of the **Enter Payroll Information** window, click on the owner's name to get to the Preview Paycheck window.

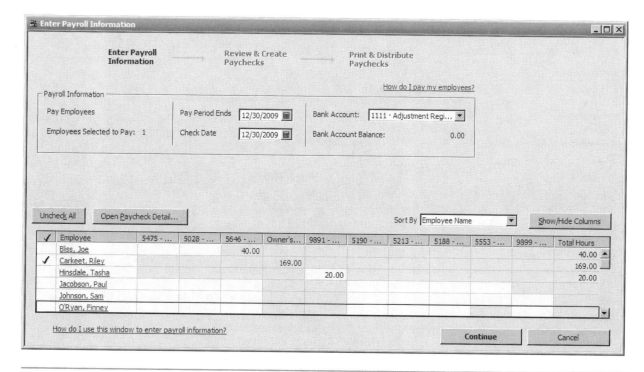

Figure 15-11

Use your Adjustment Register account when you process an owner's paycheck.

Now you'll see the Preview Paycheck window for the owner's timesheet. Figure 15-12 shows our example for Riley Carkeet. You want to be sure you get Owner's Labor Deduction in the Other Payroll Items box in this window. To do this:

▌ In the **Item Name** column, under the **Other Payroll Items** box, select the payroll item **Owner's Labor Deduction**.

▌ In the **Rate** column enter –100%.

▌ Make sure **Check Amount** is 0.0 when you finish entering these amounts.

▌ Click **Save & Close.**

▌ Click **Continue** in the Enter Payroll Information window.

▌ In the Review and Create Paychecks window, click **Create Paychecks**.

▌ Click **Close.**

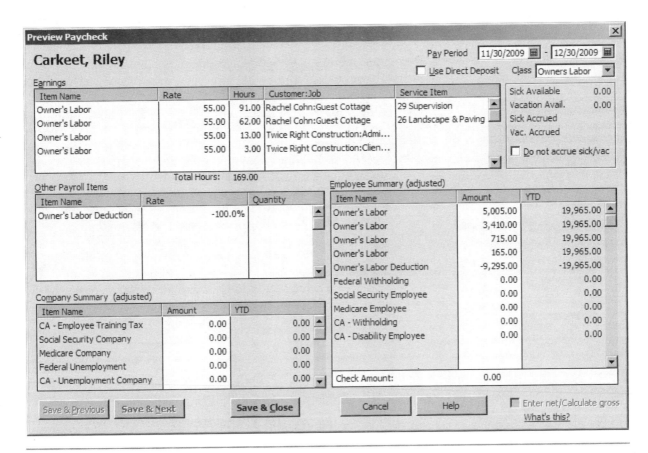

Figure 15-12
When processing payroll for an owner, make sure the Owner's Labor Deduction and rate are entered correctly. It's important that the Check Amount is 0.00.

In this chapter, we gave you a great deal of information. We told you how to:

▌ enter timesheets

▌ process payroll

▌ allocate workers' comp costs to jobs

▌ get owner's time into job cost reports

If you feel a bit overwhelmed by all the information we've presented, keep in mind that you don't have to take all of this in at one time. For example, you can simply process payroll first and then add workers' comp costs later. Then you can add owner's time when you have payroll and workers' comp figured out. In other words, you can take things one step at a time.

Chapter 16

Using QuickBooks Pro on a Cash Basis

You can use QuickBooks Pro on a cash basis or an accrual basis. Or, you can start on a cash basis and later change to an accrual basis. All you have to do is change one option in your preferences.

In the previous two chapters we showed you how to do your receivables and payables on the accrual basis. That's what most contractors will use. But if you're a small operator with few or no employees, or no time to do bookkeeping, you're probably running your business on a cash basis.

If you're just starting your business or have just made the transition from Quicken, you may feel more comfortable starting to use QuickBooks Pro on a cash basis. Although this may be an easier way to get up and running, we suggest that you try to change to the accrual method as soon as possible. Accrual-based reports will give you a more timely and accurate picture of your finances, job status, and operating status. Since QuickBooks Pro has the ability to switch reporting between cash and accrual, your accountant can still file your tax returns on a cash basis, and you can get your management reports on the more accurate accrual basis.

You may decide to use the cash basis of accounting if:

▌ you're just starting your business and know "off the top of your head" what the payables and receivables are.

▌ you've just started to get computerized and are overwhelmed by getting everything into the computer.

▌ you have very little time to do bookkeeping.

So let's look at the cash basis of accounting. To report an expense, you simply write a check and log it in. To report income, you deposit money in the bank and log in the deposit. It's easy and fast. You don't have to enter a bill or invoice and then tell the computer it has been paid. The downside, of course, is that the computer isn't tracking who you owe money to, or who owes you money. Also, your reports don't include expenses that you haven't paid or invoices you haven't issued to customers.

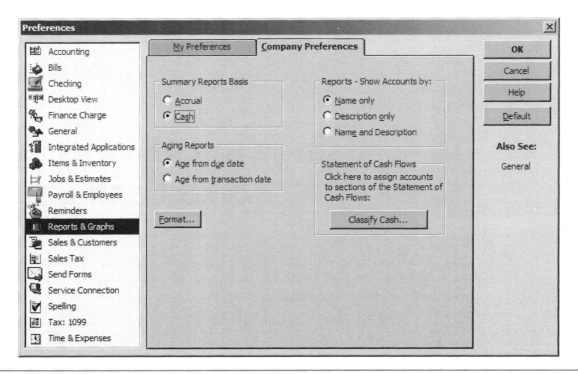

Figure 16-1
Use the Reports & Graphs Preferences window to let QuickBooks Pro know you operate on a cash basis.

To tell QuickBooks Pro that you want to operate on a cash basis:

▌ From the **Edit** menu, choose **Preferences**.

▌ Click the **Reports & Graphs** icon. See Figure 16-1.

▌ Click the **Company Preferences** tab.

▌ Select **Cash** for the Summary Reports Basis.

▌ Click **OK**.

How to Record a Check

To record a check:

▌ Click **Home**, then click **Write Checks**, or from the **Banking** menu, choose **Write Checks**.

Now let's look at Figure 16-2 to see what you would need to enter here:

▌ In **Bank Account**, make sure you select the correct bank account.

▌ Check **To be printed** unless you're recording a handwritten check.

▌ In **No.**, enter the check number if it's a handwritten check.

Figure 16-2

The QuickBooks Pro Write Checks window looks similar to a regular paper check.

▌ In **Date**, enter the date of the check.

▌ In **Pay to the Order of**, enter payee name or select the name from the drop-down vendor list.

▌ In **$**, enter the check amount. The Dollars field will fill in automatically.

Using the Items Tab to Record a Job-Related Cost

You should always use the Items tab to enter job-related costs. Our example in Figure 16-3 shows the Items tab on a check to Pella Windows for the job-related cost of windows for Jennifer Blanco:Kitchen Remodel.

To record a check for a job-related cost in a Write Checks window:

▌ Click the **Items** tab. Click in the **Item** column and select the appropriate items from the drop-down list.

▌ In **Qty**, enter the quantity of the item. Amount will be calculated automatically.

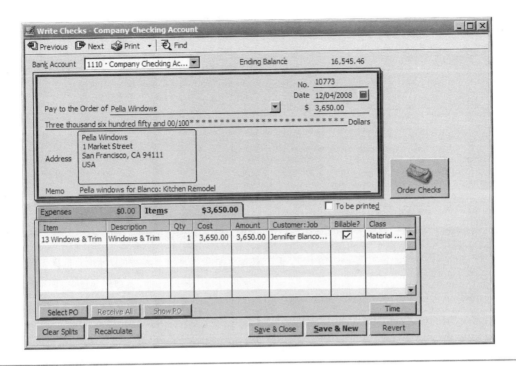

Figure 16-3
It's important to use the Items tab when you write checks for job-related expenses.

▌ Select **Customer:Job** from the pull-down Customer:Job list.

▌ Put a check in the Billable column if you are billing the item amount to your client.

▌ In **Class**, select the appropriate class from the pull-down Class list (if you're using classes).

If you're splitting the total payment to the vendor over more than one job (or if the payment covers more items), click in the next line and make the additional entry(s) using the same procedure.

▌ Click **Save & Close**.

Using the Expenses Tab to Record an Overhead Expense

You should always use the Expenses tab to enter overhead costs. Our example in Figure 16-4 shows check number 10774 to Acme Fast Fuel for gasoline, which is an overhead expense.

To enter an overhead expense check in a Write Checks window:

▌ Click the **Expenses** tab.

▌ Click in the Account column and select the correct expense account from the drop-down list. In our example, we used 6101 - Gas & Oil.

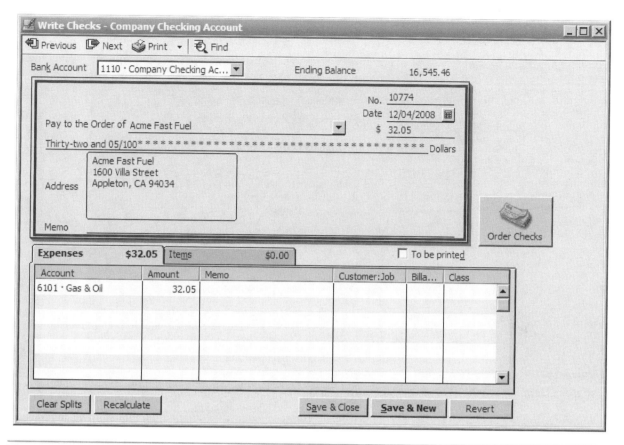

Figure 16-4
Use the Expenses tab when you write checks for overhead expenses.

If you're splitting the total payment to the vendor over more than one expense account, click in the next line and make the additional entry(s) using the same procedure.

When the check entry is complete:

■ Click **Save & Close** to record the check.

How to Record a Deposit

You record a deposit to your checking account in the Make Deposits window. To record a deposit:

■ Click **Home**, then click the **Record Deposits** icon, or from the **Banking** menu, choose **Make Deposits**.

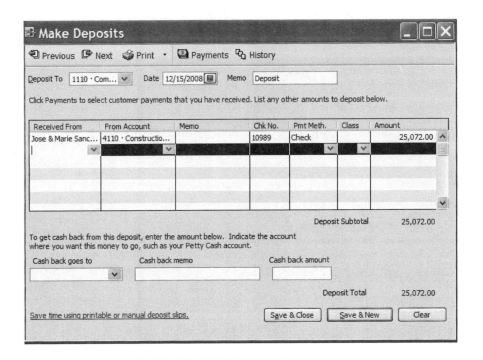

Figure 16-5
The Make Deposits window shows all checks that are ready to be deposited.

Look at Figure 16-5 to see what you would need to enter here:

▌ In the Make Deposits window, in **Deposit To**, select your checking account from the pull-down list. If you have multiple checking accounts, make sure you select the correct one.

▌ In **Date**, enter the date of the deposit.

▌ In **Received From**, use the pull-down list to select who you've received the check from.

▌ In **From Account**, select the correct account for the deposit from the pull-down list. Usually this is an income account.

▌ In **Chk No.**, enter the check number.

▌ In **Pmt Meth.**, select payment method from the pull-down list.

▌ In **Class**, select the correct class for the deposit from the pull-down list. Usually, you'll use Revenue - Job Related. If the deposit isn't job related, you can skip this.

▌ In **Amount**, enter the deposit amount.

▌ Click **Save & Close** to record the deposit.

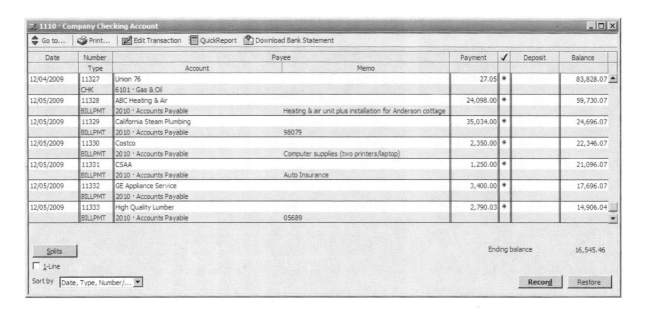

Figure 16-6
Your checking account register shows all your payments, deposits, and your account balance.

Checking Your Transactions with the QuickBooks Pro Register

QuickBooks Pro has an on-screen register you can use to look at all your recorded transactions. It displays all of your recorded transactions in a list. For example, you can use the register to get a quick look at all the checks you've written over a certain time. Or you can use it to get your checkbook account balance.

To open the Checking Register:

 ▌ Click **Home**, then click **Chart of Accounts**, or from the **Lists** menu, choose **Chart of Accounts**.

 ▌ In the Chart of Accounts list, double-click your checking account to bring up the register. See Figure 16-6.

This register will show you payments, deposits, and current balance of your checking account.

Now that you've learned how to enter transactions on a cash basis, keep in mind that running your construction company on a cash basis isn't the best way to run the business. As soon as possible, you should switch to an accrual basis of accounting so your job cost reports will always be up to date.

Reports

As a contractor, you need up-to-the-minute information on job costs — but "up-to-the-minute" means different things to different contractors. If you typically run jobs that last only a week or so, you'll probably want to analyze your job costs daily. That's the only way to keep up with them. If your jobs run months or years, you don't need to get job cost reports so frequently. Once a month may be sufficient for longer jobs.

In this chapter, we'll look at reports for the construction business on job costing. We'll cover:

▌ how to customize existing reports.

▌ how to use the memorized reports in the *sample.qbw* and *company.qbw* data files on the CD-ROM that comes with this book.

▌ how to use the project reports in QuickBooks Pro.

Finding and learning to use reports will give you more insight into the day-to-day status of your jobs. Learning how to modify reports will give you the power to create reports that are meaningful to you and more specific to your business.

We added the memorized reports in *sample.qbw* and *company.qbw* to give you reports you can use on a day-to-day basis. We've customized the reports for the construction industry.

How to Modify Reports

You can use the Modify Report window to pick the information you want to include in the report. For example, you may want to get a report for just one job, or you may want to change a report to include an invoice or check number. When you display a report on your computer, you'll see the Modify Report window blocking the report. To view the report, click the cancel button on the Modify Report window. To retrieve the Modify Report window, click the button titled Modify Report in the top left corner of the report window. You can see this button in Figure 17-1, where we show a Job Profitability Detail report.

Figure 17-1
To change a report, click the Modify Report button located in the top left corner of the report window.

Note: When Modifying reports from the Report Center, enter the date range when prompted. To make further changes to the report, click the Modify Report button.

The Display Tab on the Modify Report Window

In general, you use the Display tab of a Modify Report window to select the date range and the columns you would like to see on a report. Figure 17-2 shows the Display tab column settings for the Job Profitability Detail report in Figure 17-1. It's worth noting here that the Display tab options change depending on the report you're modifying. For example, in Figure 17-2 you can only select from the four columns shown in the Display tab window — Act. Cost, Act. Revenue, $ Difference, and % Difference. In Figure 17-3, we show the Display tab of the Modify Report window for a Customer Phone List report. There are many more options for this report.

Figure 17-2

The Display tab window lets you specify which dates and columns will be on the report.

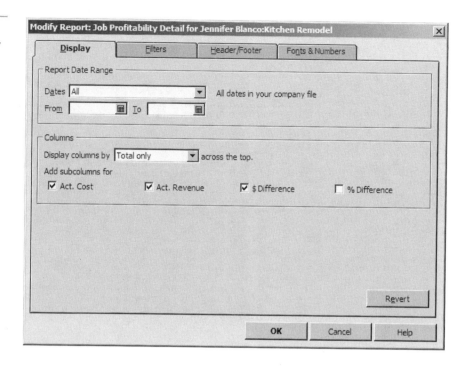

Figure 17-3

The Display tab window changes depending on the type of report you're generating.

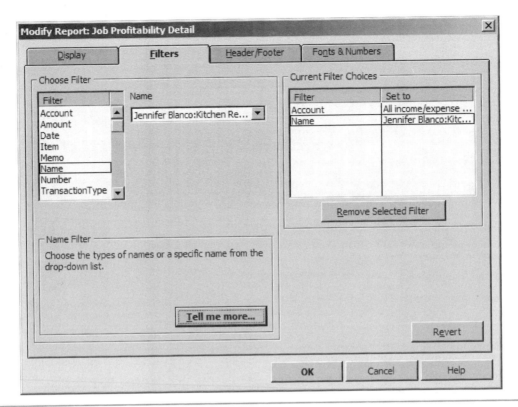

Figure 17-14
The Filters tab window lets you pick which data is on your report.

The Filters Tab on the Modify Report Window

The Filters tab on the Modify Report window lets you specify which specific records will appear in a report. You can define the accounts, customers:jobs, classes, transaction types, payees, dates, transaction amounts, and/or memo contents that each record must match before it's included in the report. In Figure 17-4, we show a Filters tab window to get a report for a particular job — Jennifer Blanco:Kitchen Remodel. At the right side of the Filters tab window under Current Filter Choices, you'll see a list of the current filters for the report.

The Header/Footer Tab on the Modify Report Window

The Header/Footer tab on the Modify Report window lets you customize your report and modify your header and footer information. When memorizing a new report, you'll use this tab to change the report's title.

Figure 17-5

The Memorized Report List in *sample.qbw* and *company.qbw*.

Using Our Memorized Reports

Memorizing a report is easy — but creating the report the first time may take some time. Determining exactly what combination of modifying and filtering will give you the report you want can be a bit of a brain twister. Also, you have to be absolutely consistent and accurate when you enter transactions. If you enter a record incorrectly, it may not be in the reports when, or where, it should be. You don't want to make a good decision based on bad information.

We defined and memorized the reports we discuss in this section in the *company.qbw* and *sample.qbw* data files included on the CD that comes with this book. We hope you can make good use of these reports in managing your business. We based the reports on the *sample.qbw* and *company.qbw* data files. If you've changed these files or are using other data files, you'll need to review the reports to make sure they show what you need.

However, you may have to change the dates used in the memorized reports. You can use the methods discussed in the previous sections to do that.

To display a list of the memorized reports in *sample.qbw* or *company.qbw*:

▌ Open **sample.qbw** or **company.qbw**.

▌ Click **Report Center**, then **Memorized**, or from the **Reports** menu, choose **Memorized Reports** and then **Memorized Report List**. See Figure 17-5.

▌ To generate a memorized report, select the report and click **Display**.

Figure 17-6
A memorized Chart of Accounts Report.

If you didn't use the *sample.qbw* or *company.qbw* data file to set up your QuickBooks Pro company file, you may want to memorize some of the reports in this section in your own company file. We'll tell you how to do that after we describe each report.

However, if you did use our *sample.qbw* or *company.qbw* data file to set up your company data file, you can just skip the information on memorizing each report.

Now let's look at each memorized report.

Chart of Accounts Report

The memorized Chart of Accounts Report simplifies the QuickBooks Pro built-in Chart of Accounts Report. It looks similar to the one we printed and included in Chapter 3. Use this report to help define which accounts you have set up in your Chart of Accounts. Figure 17-6 shows a memorized Chart of Accounts Report.

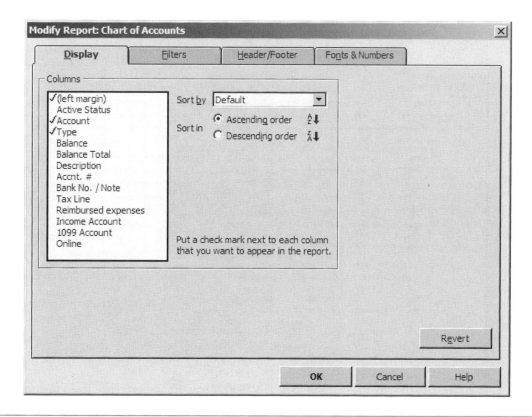

Figure 17-7
Choose the columns to appear in the Chart of Accounts Report.

Memorizing the Chart of Accounts Report — To memorize the Chart of Accounts Report in your QuickBooks Pro data file:

▌ Click **Report Center**, or from the **Reports** menu, choose **List**, then **Account Listing**.

▌ Now select the columns in the Display tab of the Modify Report window, as shown in Figure 17-7. Select only **(left margin)**, **Account**, and **Type** from the Columns list. Be sure to scroll down the list to deselect other columns on the list.

▌ Click the **Header/Footer** tab. In **Report Title**, re-type **Chart of Accounts Report**.

▌ Click **OK**.

▌ In the report window, click **Memorize**.

▌ In the **Memorize Report** window, enter **Chart of Accounts**.

▌ Click **OK**.

Figure 17-8
A memorized Job Cost Class Report.

Job Cost Class Report

The memorized Job Cost Class Report shows you, in detail, all revenue and costs separated by class. This report is a Transaction Detail by Account report sorted by these classes — Revenue, Labor, Materials, Subcontractor, Equipment Rental, and Other. The bold column header at the top left of each section is the class. In the example we show in Figure 17-8, it's Material Costs - Job Related.

Memorizing the Job Cost Class Report — To memorize the Job Cost Class Report in your QuickBooks Pro data file:

▌ Click **Report Center**, or from the **Reports** menu, choose **Accountant & Taxes**, and then **Transaction Detail by Account**.

▌ Select the columns in the Display tab of the Modify Report window, as shown in Figure 17-9. Select only **(left margin), Type, Date, Num, Source Name, Amount**, and **Balance**. Scroll down to the bottom of the list to find Source Name, Amount, and Balance.

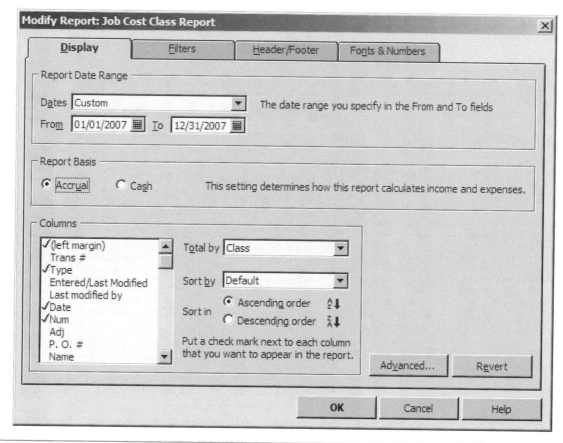

Figure 17-9
In the Total by field, select Class to group and subtotal transactions by job cost class.

∎ In **Report Date Range**, enter the date range you're interested in. Usually this would be for a given month or year. Our example in Figure 17-9 uses the fiscal year 1/1/2007 to 12/31/2007.

∎ In **Total By**, select **Class** from the pull-down list, then click **OK**.

∎ In the report window, click **Memorize**.

∎ In the **Memorize Report** window, enter **Job Cost Class Report**.

∎ Click **OK**.

Often you'll want to know the job-related labor costs for just one job. To do this:

∎ Click the **Modify Report** button when you have a Job Cost Class report on your screen.

∎ Click the **Filters** tab.

Figure 17-10

To filter a report for a specific job, select the job from the Name drop-down list.

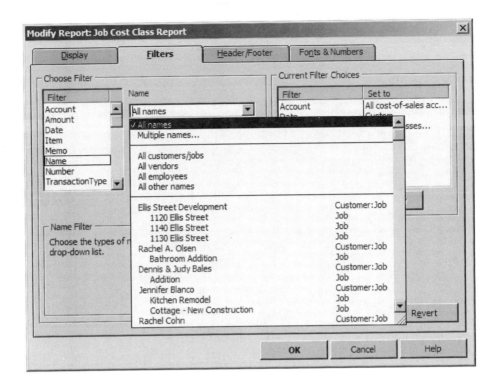

▌ In the Filters tab window, select **Name** from the Filter list. See Figure 17-10.

▌ From the Name drop-down list, select the **Customer:Job** you're interested in.

▌ Click **OK**.

Retentions Receivable Report

The Retentions Receivable Report lists all the customers that owe you for retention for the dates you specify. See Figure 17-11.

Memorizing the Retentions Receivable Report — To memorize the Retentions Receivable Report in your QuickBooks Pro data file:

▌ Click **Report Center**, or from the **Reports** menu, choose **Customers & Receivables**, and then **Customer Balance Summary**.

▌ In **Report Date Range**, enter the date range you're interested in. Usually this would be for a given month or year.

▌ Click the **Filters** tab.

Figure 17-11

A memorized Retentions Receivable Report.

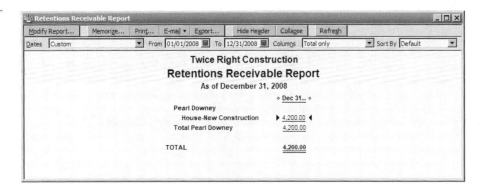

Figure 17-12

The Report Filters window showing some of the settings for the Retentions Receivable Report.

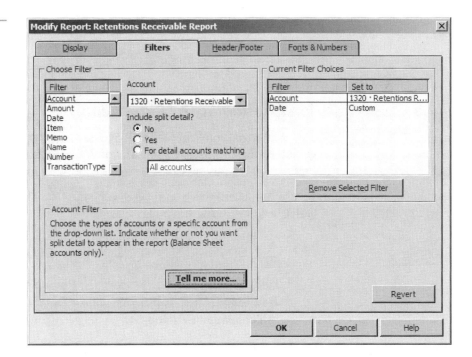

▌ In the Filters tab window, select **Account** from the Filter list.

▌ From the pull-down Account list, select your **Retentions Receivable** account. In our example shown in Figure 17-12, the account is 1320 - Retentions Receivable.

▌ Click **OK**.

▌ In the report window, click **Memorize**.

▌ In the **Memorize Report** window, enter **Retentions Receivable Report**.

▌ Click **OK**.

Figure 17-13

A memorized Job Profitability by Class report.

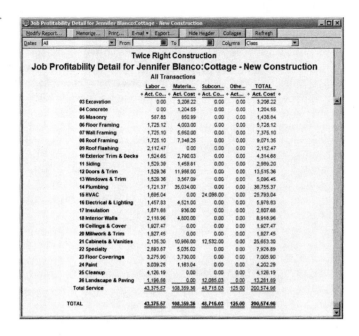

Job Profitability by Class Report

The memorized Job Profitability by Class report breaks down all costs for a single job by these classes: Labor, Material, Subcontractor, Equipment, and Other. Figure 17-13 shows an example of a memorized Job Profitability Detail report.

To use the *Memorizing the Job Profitability by Class Report* feature in your QuickBooks Pro data file:

▌ Memorizing the Job Profitability by Class Report - To memorize the Job Profitability by Class Report in your QuickBooks Pro data file:

▌ Click **Report Center**, or from the **Reports** menu, choose **Jobs, Time & Mileage**, then **Job Profitability Detail**. You will be prompted to select a Customer:Job.

▌ Set the columns area in the Display tab of the Modify Report window as shown in Figure 17-14.

▌ Click on the Filters tab. Select **Class** from the Filter list.

▌ From the pull-down Class list, select **Multiple classes** to bring up the pop-up window showing all your classes, as shown in Figure 17-15.

▌ Select the classes Labor - Job Related, Material Costs - Job Related, Subcontractor Costs - Job Related, Equip Rental Costs - Job Related, and Other Job Costs. Your class names may be slightly different.

Figure 17-14

In the Columns section, select Display columns by Class.

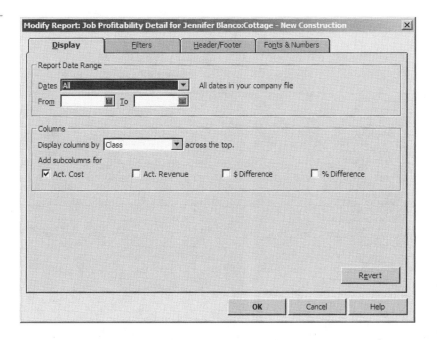

Figure 17-15

Filter your report to include only the classes you want displayed.

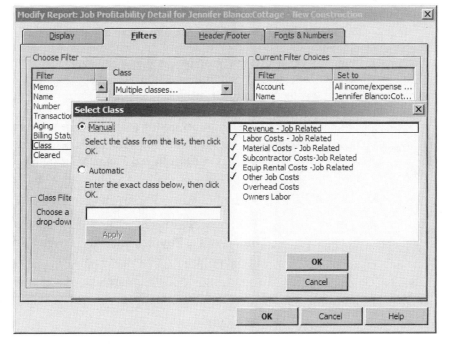

▌ Click **OK** to close the Select Class window.

▌ Click on **OK** to display the report.

▌ Click **Memorize** in the report window.

▌ In the Memorize Report window, enter Job Profitability by Class.

▌ Click **OK**.

Twice Right Construction
Workers' Comp Report
December 2007

	Bliss, Joe			Hinsdale, Tasha			Jacobson, Paul			Johnson, Sam			O'Ryan, Finney			TOTAL	
	Hours	Rate	Dec 07	Hours	Rate	Dec 07	Hours	Rate	Dec 07	Hours	Rate	Dec 07	Hours	Rate	Dec 07	Hours	Dec 07
Employee Wages, Taxes and...																	
Gross Pay																	
5190 - Electrical			0.00			0.00			0.00			0.00	8	21.25	170.00	8	170.00
5213 - Concrete			0.00			0.00	72	22.00	1,584.00	32	21.25	680.00	64	21.25	1,360.00	168	3,624.00
5475 - Painting			0.00			0.00			0.00			0.00	32	21.25	680.00	32	680.00
5646 - Carpentry	40	21.25	850.00			0.00	8	22.00	176.00	48	22.00	1,056.00	192	21.25	4,080.00	288	6,162.00
9891 - Office Labor			0.00	20	15.00	300.00			0.00			0.00			0.00	20	300.00
Total Gross Pay	40		850.00	20		300.00	80		1,760.00	80		1,736.00	296		6,290.00	516	10,936.00
Adjusted Gross Pay	40		850.00	20		300.00	80		1,760.00	80		1,736.00	296		6,290.00	516	10,936.00
Taxes Withheld																	
Federal Withholding			-156.00			-40.00			-250.00			-189.00			-782.00		-1,417.00
Medicare Employee			-12.33			-4.35			-25.52			-27.41			-87.58		-157.19
Social Security Employee			-65.03			-22.95			-138.92			-166.46			-448.76		-840.12
CA - Withholding			0.00			0.00			-46.00			-120.00			-208.00		-374.00
CA - Disability Employee			0.00			0.00			0.00			-35.00			0.00		-35.00
Total Taxes Withheld			-233.36			-67.30			-460.44			-537.87			-1,524.34		-2,823.31
Deductions from Net Pay																	
Medical deduction			-50.00			0.00			0.00			-50.00			-200.00		-300.00
Total Deductions from Net...			-50.00			0.00			0.00			-50.00			-200.00		-300.00
Net Pay	40		566.64	20		232.70	80		1,299.56	80		1,148.13	296		4,565.66	516	7,812.69

Figure 17-16

A Payroll Summary by Employee report, customized to report on workers' compensation.

Workers' Compensation Report

The memorized Workers' Compensation Report lists the allocated payroll transactions by workers' compensation category. However, you must have used workers' comp codes as payroll items to use this report. The totals shown will be for the report dates you select. Figure 17-16 shows a memorized Workers' Comp (titled Payroll Summary) Report.

Memorizing the Workers' Compensation Report — To memorize the Workers' Compensation Report in your QuickBooks Pro data file:

▌ Click **Report Center**, or from the **Reports** menu, choose **Employees & Payroll**, and then **Payroll Summary**.

▌ On the Display tab of the Modify Report window, change the date range of the report to match the dates your Workers' Compensation Insurance carrier asks for, as shown in Figure 17-17.

▌ Click **OK**.

▌ Click **Memorize** in the report window.

▌ In the Memorize Report window, enter **Workers' Comp Report**.

▌ Click **OK**.

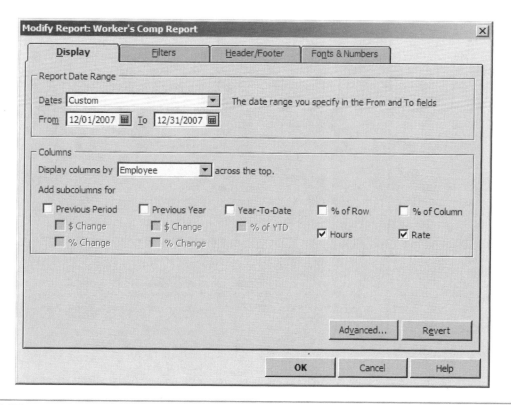

Figure 17-17
Set the dates for the Workers' Comp Report in this window.

Job Costs Detail Report

The memorized Job Costs Detail report lists and totals all job cost transactions for any job you select, and summarizes the transactions by item or job phase. Figure 17-18 shows an example of a memorized Job Costs Detail report.

Memorizing the Job Costs Detail Report — To memorize the Job Costs Detail report in your QuickBooks Pro data file:

▌ Click **Report Center**, or from the **Reports** menu, choose **Accountant & Taxes**, and then **Transaction Detail by Account**.

▌ From the pull-down Dates list, select **All**.

▌ Now select the columns in the Display tab of the Modify Report window, as shown in Figure 17-19. Select only **(left margin)**, **Type**, **Date**, **Num**, **Source Name**, **Debit**, and **Credit**. Scroll down to the bottom of the list to find Source Name, Debit, and Credit.

▌ From the **Total by** pull-down list, select **Item detail**.

▌ Click the **Filters** tab.

Figure 17-18

A memorized Job Costs Detail report.

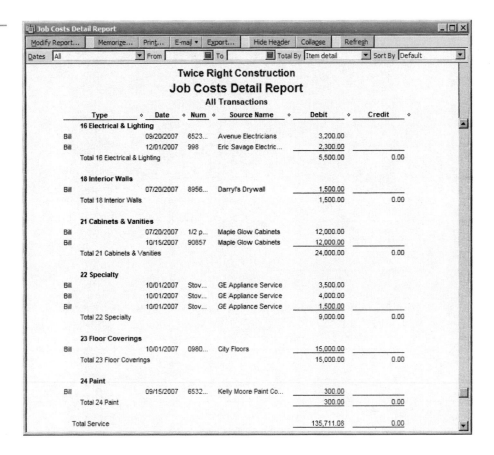

Figure 17-19

Choose the columns to appear in the Job Costs Detail report.

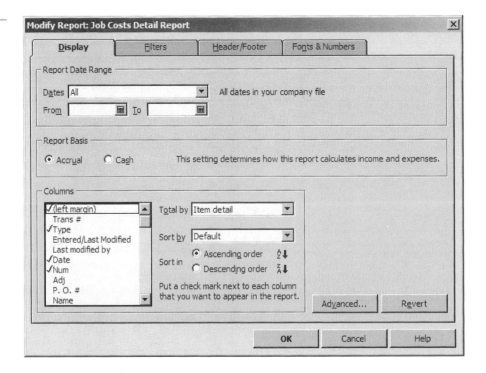

Figure 17-20

Choose the transaction types to include in the Job Costs Detail report.

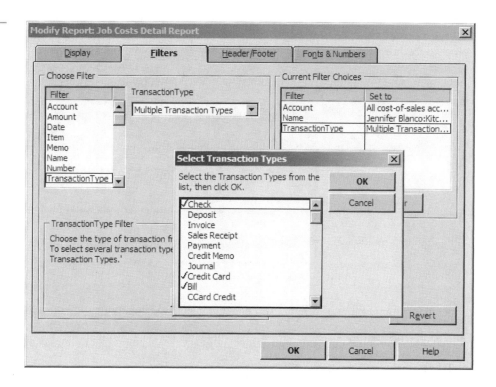

I Under **Filter**, select **Account**.

I From the pull-down Account list, select **All cost-of-sales accounts**.

I Under **Filter**, select **Transaction Type**. See Figure 17-20.

I From the pull-down Transaction Type, select **Multiple Transaction Types** and then select **Check**, **Credit Card**, **Bill**, **CCard Refund**, **Bill Credit**, **Bill CCard**, **Item Receipt**, and **Paycheck**.

I Click **OK** to close the Select Transaction Type window.

I Under **Filter**, select **Name**. In the **Name** drop-down list, select the job you wish to report on.

I Click the **Header/Footer** tab, and in the **Report Title**, change to **Job Costs Detail Report**.

I Click **OK**.

I Click **Memorize** in the report window.

I In the Memorize Report window, enter **Job Costs Detail**.

I Click **OK**.

Figure 17-21

A memorized Unpaid Job Bills report.

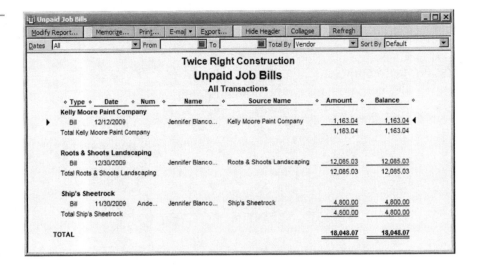

Figure 17-22

Choose the columns to appear in the Unpaid Job Bills report.

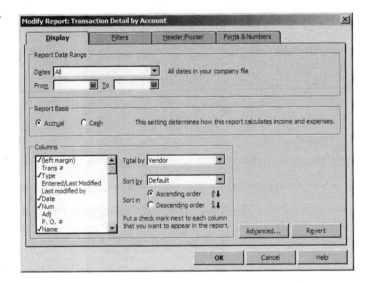

Unpaid Job Bills

This report will list and total all unpaid job expenses. Transactions are grouped and subtotaled by vendor, which are the bold headings on the report. Figure 17-21 shows an example of a memorized Unpaid Job Bills report.

Memorizing the Unpaid Job Bills Report — To memorize the Unpaid Job Bills report in your QuickBooks Pro data file:

▌ Click **Report Center**, or from the **Reports** menu, choose **Accountant & Taxes**, and then **Transaction Detail by Account**.

▌ Now select the columns in the Display tab of the Modify Report window, as shown in Figure 17-22. Select only **(left margin)**, **Type**, **Date**, **Num**, **Name**, **Source Name**, **Amount**, and **Balance**. Scroll down to the bottom of the list to find Source Name, Amount, and Balance.

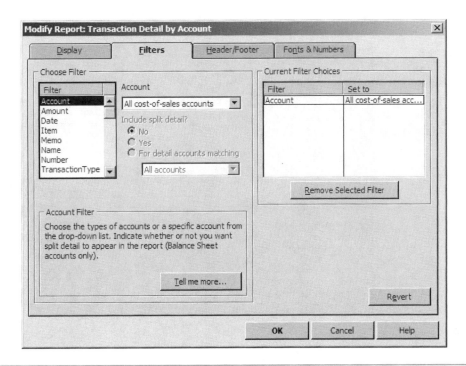

Figure 17-23
Choose the accounts to include in the Unpaid Job Bills report.

▮ In **Dates**, select **All**.

▮ In **Total by**, select **Vendor** from the pull-down list.

▮ Click the **Filters** tab.

▮ Under **Filter**, select **Account** and then from the pull-down Account list, select **All cost-of-sales accounts**. See Figure 17-23.

▮ Under **Filter**, select **Transaction Type** and then from the pull-down Transaction Type list, select **Bill**.

▮ Under **Filter**, select **Paid Status** and then **Open**, as shown in Figure 17-24.

▮ Click the **Header/Footer** tab. In the **Report Title** change to **Unpaid Job Bills Report**.

▮ Click **OK**.

▮ Click **Memorize** in the report window.

▮ In the Memorize Report window, enter **Unpaid Job Bills**.

▮ Click **OK**.

Figure 17-24

Choose the status of the transactions to include in the Unpaid Job Bills report.

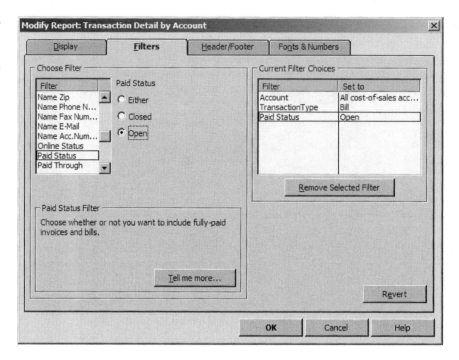

Vendors' WC Expiration Dates

The memorized Vendor Workers' Comp Expiration Dates report lists all subcontractors, with the phone, contact name, and workers' comp expiration date. If you've made an entry into the custom field for "workers' comp expires," that date will display in the far right column. Refer back to Chapter 14 for more information on creating and using the "comp expires" custom field. Figure 17-25 shows a memorized Vendors' WC Expiration Dates report.

Memorizing the Vendors' WC Expiration Dates Report — To memorize the Vendor's WC Expiration Dates report in your QuickBooks Pro data file:

▌ Click **Report Center**, or from the **Reports** menu, choose **Vendors & Payables** and **Vendor Contact List**.

▌ Now select the columns in the Display tab of the Modify Report window, as shown in Figure 17-26. Select only **(left margin)**, **Vendor**, **Phone**, **Contact**, and **Workers' Comp Expires**. We've scrolled down to the bottom of the list to show Workers' Comp Expires in the list, but it won't be there if you didn't set up a custom field for it.

▌ Click the **Header/Footer** tab. In **Report Title**, type **Vendors' WC Expiration Dates.**

▌ Click **OK**.

▌ Click **Memorize** in the report window.

▌ In the Memorize Report window, enter **Vendors' WC Expiration Dates**.

▌ Click **OK**.

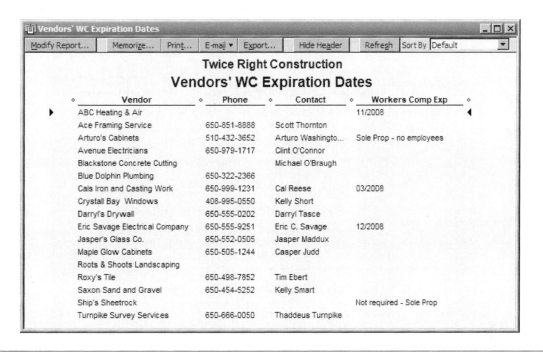

Figure 17-25
A memorized Vendors' WC Expiration Dates report.

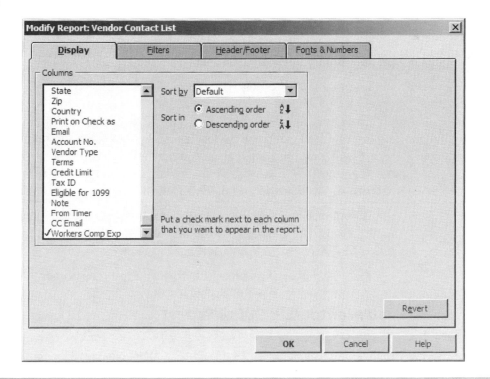

Figure 17-26
Choose the columns to appear in the Vendors' WC Expiration Dates report.

Using Jobs, Time & Mileage Reports

You can use the built-in QuickBooks Pro Jobs, Time & Mileage project reports to help you bill customers, analyze how well each job is progressing financially, and identify problem jobs before it's too late. We'll discuss what each report shows, how to use the report in your business, and when to run it. We'll also show you a sample printout of each report. We didn't discuss the Item Profitability, Item Estimates vs. Actuals, Profit & Loss by Job, or Time by Item reports because they aren't useful to contractors.

To run a project report:

▌ Click **Report Center**, or from the **Reports** menu, choose **Jobs, Time & Mileage** and select the report by name.

When the Modify Report window appears, you can usually accept the default values and just click OK to create the report. In some cases you'll want to specify a time period for the report.

When running certain reports, such as the Job Profitability Detail report, you'll be asked to select the job you want to run the report for in the Filter Report by Job window.

Job Profitability Summary Report

The Job Profitability Summary report lists every job you're working on for each of your customers. This report lists the jobs by name, actual job costs to date, actual revenue or total amount of billings to date, and in the last column, dollar difference. Dollar difference is the actual costs to date minus the total amount billed to the job in dollars. Figure 17-27 shows an example of a Job Profitability Summary.

Use this report to help you bill customers. Since the report gives you a good idea of where each project's actual costs are in relation to its billings, you can use it to get an idea of how much you should bill on each job. For example, if you bill your customers each week on a lump sum basis, this report will help you analyze exactly what the costs are to date and how much you've billed so far. Use this information to bill each job so your costs and profits are covered and the job is at or below its estimated cost.

If you've billed 100 percent of a job, the report will tell you how profitable it was. For example, if the actual cost of a job is $15,000 and you billed $17,000, the $2,000 difference is profit.

Print (and check) this report before you bill a customer.

Job Profitability Detail Report

The Job Profitability Detail report shows one job, broken down by job phases (items). It tells you the actual job costs to date, actual revenue or total amount of billings to date, and in the last column, the dollar difference. Dollar difference is the actual costs to date minus the total amount billed for each job phase, in dollars. Figure 17-28 shows a sample Job Profitability Detail report.

Figure 17-27

Use the Job Profitability Summary report to help you when you invoice customers.

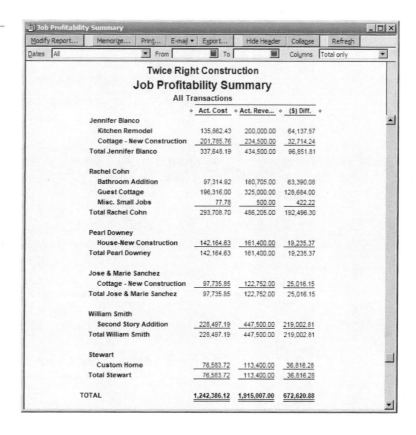

Figure 17-28

The Job Profitability Detail report breaks down a particular job by item.

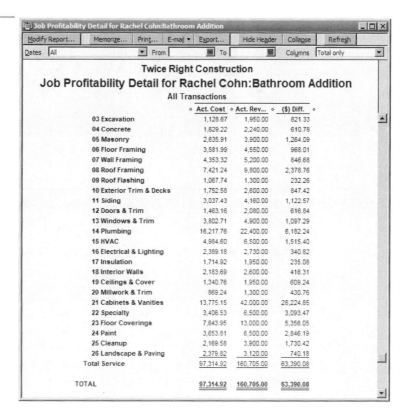

Figure 17-29

Print this report often to check actual vs. estimated costs.

Twice Right Construction

Job Estimates vs. Actuals Summary

All Transactions

	Est. Cost	Act. Cost	($) Diff.	Est. Reve...	Act. Reve...	($) Diff.
Dennis & Judy Bales						
Addition	120,000.00	66,047.84	-53,952.16	120,000.00	144,000.00	24,000.00
Total Dennis & Judy Bales	120,000.00	66,047.84	-53,952.16	120,000.00	144,000.00	24,000.00
Jennifer Blanco						
Kitchen Remodel	209,150.00	136,389.43	-72,760.57	209,395.56	205,200.00	-4,195.56
Cottage - New Construction	579,590.00	214,742.76	-364,847.24	579,590.00	234,500.00	-345,090.00
Total Jennifer Blanco	788,740.00	351,132.19	-437,607.81	788,985.56	439,700.00	-349,285.56
Rachel Cohn						
Bathroom Addition	119,150.00	97,314.92	-21,835.08	160,705.00	160,705.00	0.00
Guest Cottage	205,300.00	206,106.00	806.00	266,890.00	325,000.00	58,110.00
Total Rachel Cohn	324,450.00	303,420.92	-21,029.08	427,595.00	485,705.00	58,110.00
Pearl Downey						
House-New Construction	138,455.00	142,164.63	3,709.63	207,682.50	157,200.00	-50,482.50
Total Pearl Downey	138,455.00	142,164.63	3,709.63	207,682.50	157,200.00	-50,482.50
William Smith						
Second Story Addition	758,700.00	228,497.19	-530,202.81	900,750.00	447,500.00	-453,250.00
Total William Smith	758,700.00	228,497.19	-530,202.81	900,750.00	447,500.00	-453,250.00
Stewart						
Custom Home	75,600.00	76,583.72	983.72	113,400.00	113,400.00	0.00
Total Stewart	75,600.00	76,583.72	983.72	113,400.00	113,400.00	0.00
TOTAL	2,205,945.00	1,167,846.49	-1,038,098.51	2,558,413.06	1,787,505.00	-770,908.06

Use this report to help you bill a job on a percentage-complete basis by phase. Since this report gives you a good idea of where each project's actual costs are in relation to its billings, it'll help you decide how much you should bill by phase. For example, if you bill a job based on how much work has been completed on each phase of construction, use this report to analyze exactly what the costs are to date by job phase and how much you've previously billed. Then you can quickly produce a billing for the job by phase of construction.

If you've billed 100 percent of a project, the report will give you an idea of each job's profitability by phase. For example, if phase 01 - Plans and Permits actual costs is $1,992.93 and its actual revenue is $2,275, then the profit is $282.07.

Print this report before you bill a customer. It'll help you make sure a job's actual costs aren't running too high or low in relation to its billings.

Job Estimates vs. Actuals Summary Report

The Job Estimates vs. Actuals Summary report tells you how each job is progressing, and which jobs are over, or under, their estimated costs. It's a handy report for quick reference, especially if you have many small jobs that last a short time. Print the report often to check actual versus estimated costs. Use this report to adjust estimated costs if you find that certain types of jobs consistently have lower profit margins. Figure 17-29 shows an example of a Job Estimates vs. Actuals Summary.

Figure 17-30

The Job Estimates vs. Actuals Detail report breaks down a particular job by item.

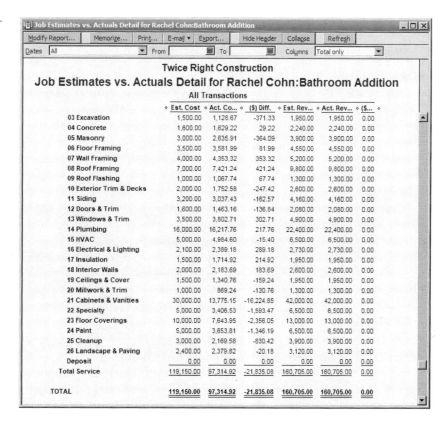

If you tend to have larger projects with many phases that span several months or years, this report may not be as useful to you as the next one, Job Estimates vs. Actuals Detail report.

Job Estimates vs. Actuals Detail Report

The Job Estimates vs. Actuals Detail report tells you, by job phase, how each job is progressing, and which part of a job is in danger of running over its estimated costs. Figure 17-30 shows a sample Job Estimates vs. Actuals Detail report.

This is a handy report for quick reference, especially if you have jobs that last a long time and you need to track them by job phase. You can also use this report to help estimate future jobs. It'll give you a good idea of which phases were over, or under, estimate.

Print this report once a week and give it to a job's Project Manager. Then he can see which part or parts of his project are over, or under, estimate while there's still time to make adjustments.

Figure 17-31

Use the Job Progress Invoices vs. Estimates report to check that the job site progress matches the percentage of progress.

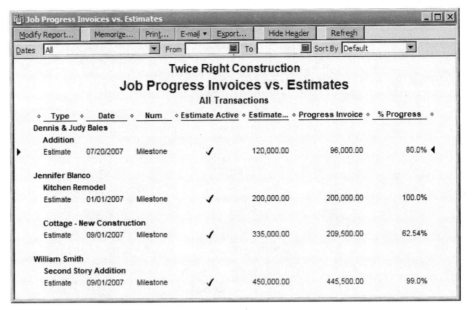

Job Progress Invoices vs. Estimates Report

The Job Progress Invoices vs. Estimates report lists each customer:job, job status, estimate total, amount invoiced to date, and percentage complete. Figure 17-31 shows a sample Job Progress Invoices vs. Estimates report.

This report gives you an idea of how much has been billed to date (in dollars) and what percentage that amount is of the total estimated costs. It's important to run this report from time to time to check whether the actual job site progress matches up with the percentage progress. For example, if you get a project status report from the field each week, compare the estimated field percentage complete to the % Progress on this report. They should be similar. If the information from the field estimates a higher percentage complete than the % Progress report, you may be behind in your billings to the customer. If the % Progress report has a higher percentage complete than the actual field progress estimate, you may have billed more than you should have in the last billing cycle.

You may need to adjust job status as work progresses. To do this:

▌ Click the **Customer Center** icon on the menu bar, or from the **Customers** menu, choose **Customer Center**.

▌ Double-click the job.

▌ Click the **Job Info** tab.

▌ In **Job Status**, select the new status from the pull-down list.

▌ Click **OK**.

Figure 17-32
An Unbilled Costs by Job report.

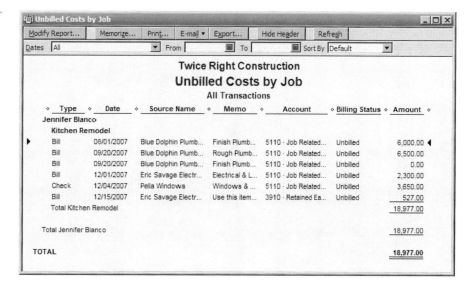

Figure 17-33
The Time by Job Summary report.

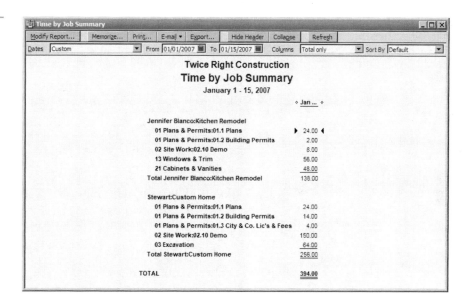

Unbilled Costs by Job Report

This report is handy if you bill your customers on a time-and-materials basis. It gives you a list of all project costs (except labor costs) that haven't been billed to a job. Use this report to double-check that you've billed all costs on a time-and-materials basis to a job. Figure 17-32 shows an example of an Unbilled Costs by Job report.

Time by Job Summary Report

This report lists each job, with detailed information for the job and phase. It's particularly useful if a job seems to be running over budget and you suspect that time has been mistakenly allocated to the wrong job or phase. Figure 17-33 shows an example of a Time by Job Summary report.

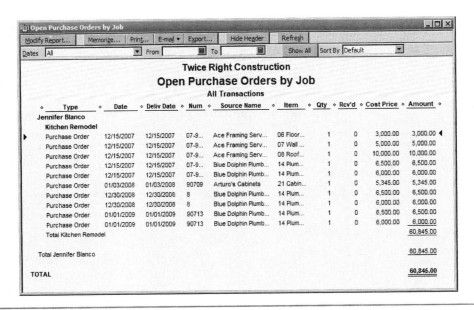

Figure 17-34
An Open Purchase Orders by Job report.

Open Purchase Orders by Job Report

In Chapter 14 we described how to use purchase orders to track committed costs or subcontractors' bids. This report will help you figure out how much you still have outstanding on those committed costs or subcontractors' bids.

QuickBooks Pro has a built-in report for tracking purchase orders that haven't been filled (in other words, a report listing all subcontractors who haven't completed or submitted bills on a project). Figure 17-34 shows an example of an Open Purchase Orders by Job report.

Important! It's worth noting here that you may have a problem with this report if you make partial payments to subcontractors. The problem occurs when you release a portion of the original PO. The remaining amount will clear from the report if you don't enter a quantity *less than 1* in the quantity field. In other words, to avoid reporting problems you should always enter an amount less than 1 when you're releasing only a portion of the PO. For more information, see Chapter 14 under the heading Using Purchase Orders to Track Multiple Draws and Committed Costs Buyouts.

The Delivery Date represents the day you were told the goods would arrive (for example, special order windows or doors) or the date the services were to start (the date the plumber or electrician said he would start work).

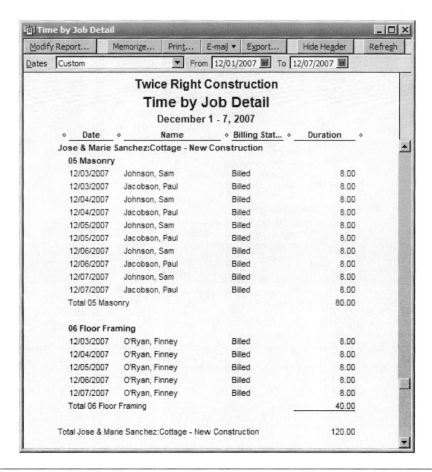

Figure 17-35
A Time by Job Detail report.

Time by Job Detail Report

This report lists each employee and his/her time on each job phase, subtotaled by job. Use this report to double-check that the time was entered correctly for each timecard, to calculate the total hours spent on a job, or to see who's spending time on specific jobs. Figure 17-35 shows an example of a Time by Job Detail report.

Time by Name Report

This report lists all employees and the hours each worked during a certain period. Use this report to find out how many hours each person has worked, and on which job, during a time period. Figure 17-36 shows part of a sample Time by Name report.

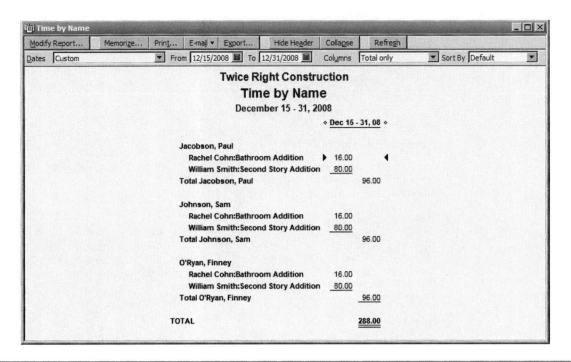

Figure 17-36
A Time by Name report.

Using Reports in QuickBooks Premier Contractor Edition

Some valuable reports are included in the Contractor Edition of QuickBooks. We've outlined two important reports below.

To run these reports:

▌ Click **Report Center**, or from the **Reports** menu, choose **Contractor Reports** and select the report by name.

Cost to Complete by Job Detail

The Cost to Complete by Job Detail report is excellent for giving you an idea of where you stand on any given job, based on the percentage complete, by phase, of the project.

▌ When running this report, you will be prompted to filter the report. Select the appropriate Customer:Job and click **OK**.

▌ Enter the percent complete for each job phase in the % Complete column. See Figure 17-37. Click **OK**.

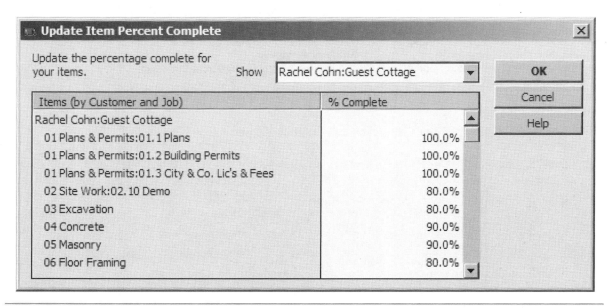

Figure 17-37
Enter the percent complete for each job phase.

▌ In the Modify Report window, click **OK** to display the report. See Figure 17-38.

Unpaid Bills by Job

The Unpaid Bills by Job report sorts all unpaid bills by job. If you pay your vendors when you receive a check from a customer and need to know which bills are outstanding for a job, you would print this report. See Figure 17-39.

In this chapter we showed you how to modify, filter, and memorize reports. We also explained the built-in QuickBooks Pro project reports.

Along with these reports, you also need to set up procedures to check your data entry. As an example, you need to be sure all your customers are being invoiced and that your Accounts Receivable are in balance. The same applies to your bills and Accounts Payable. These procedures can be broken down into end-of-month procedures and end-of-year procedures. The next chapter will walk you through those procedures, step-by-step.

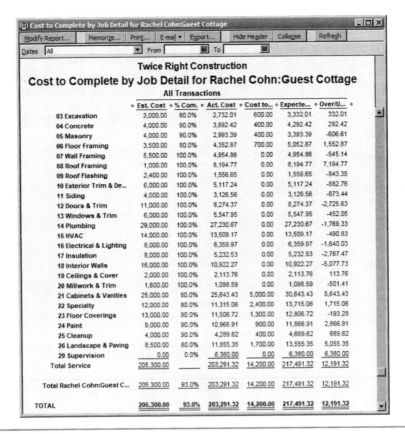

Figure 17-38

A Cost to Complete by Job Detail report.

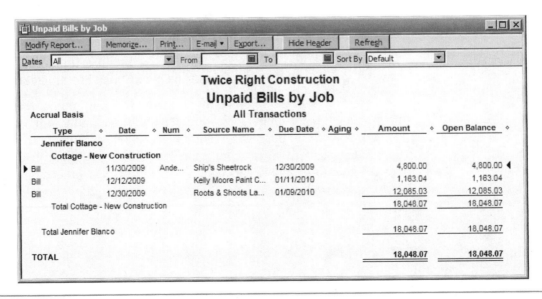

Figure 17-39

An Unpaid Bills by Job report.

Chapter 18

End of Month and End of Year Procedures

In this chapter, we'll explain what you should do at the end of each month and at the end of each fiscal year. Keep in mind that QuickBooks Pro doesn't force you to close out any month or year. We're simply suggesting what you should do to make sure that the information you've entered into QuickBooks Pro is accurate. We also suggest that you print out several reports to help you analyze how well your business is doing financially.

End of Month Procedures

Let's look first at what to do at the end of each month. You should:

- Reconcile all checking, savings, credit card, and loan accounts with your bank statement.
- Run reports to make sure all transactions are correct.
- Print monthly reports.
- Back up data and close the month.

Reconcile Each Checking Account

QuickBooks Pro makes bank account reconciliation simple, so you won't have any excuse for putting it off. Set aside the time and do it every month. Reconciling your bank account will catch any errors you or your bank have made, and serve as a check of your own accounting. Here's how to reconcile a checking account, but be sure to reconcile all savings, credit card, and loan accounts too:

- Click **Home**, then click **Reconcile**, or from the **Banking** menu, choose **Reconcile**.
- At the Begin Reconciliation window, in **Account**, select your checking account from the pull-down list. In our example shown in Figure 18-1, it's 1110 Company Checking Account.
- In **Statement Date**, enter the ending date on the bank statement.
- In **Ending Balance**, enter the amount on the bank statement (usually called New Balance or Balance this Statement).

Figure 18-1

Use the Begin Reconciliation window to select your checking account and enter all relevant information.

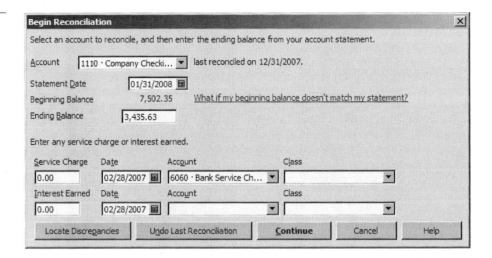

- In **Service Charge**, enter the amount listed on the statement, and the date the service charge was posted. In **Account**, select the expense account **Bank Service Charges** in your Chart of Accounts pull-down list.

- In **Interest Earned**, enter the amount listed on the statement, and the date any interest was posted. In **Account**, select the income account **Interest Income** in your Chart of Accounts pull-down list.

- Click **Continue**.

At the Reconcile window (Figure 18-2), verify each deposit, credit, check, payment, and service charge to make sure it's the same as the corresponding amount on the bank statement. If an amount is different, you have to find the problem and correct it before continuing. To make a correction, double-click the transaction and make the appropriate changes. If the bank made an error, make sure they correct it. If you find an amount on the bank statement that isn't in your company data file, it may mean you forgot to enter it. Enter it now before continuing with the reconciliation.

When your records match the bank statement, Ending Balance and Cleared Balance will be the same, and Difference will be zero.

- Click **Reconcile Now** to complete the reconciliation.

Now you'll get a dialog box asking what type of reconciliation report you would like to print:

- Select **Detail**.

- Click **Print**.

We suggest you file the printed reconciliation report and bank statement together, or set up a three-ring binder to store the reports.

Figure 18-2

Use the Reconcile window to clear checks and deposits according to your bank statement.

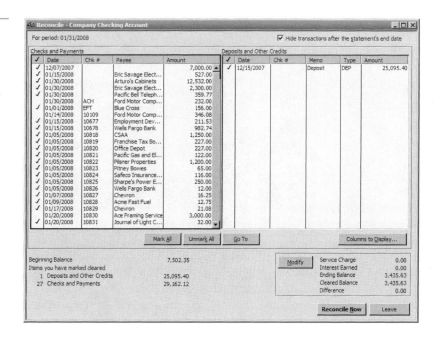

Run Reports to Make Sure All Transactions Are Correct

The three reports you should run to make sure you've entered all transactions correctly are:

▌ Transaction Detail by Account

▌ A/R Aging Detail

▌ A/P Aging Detail

On each report, check each transaction to make sure it's correct and posted to the correct account. You have to correct any mistakes before continuing. While you have each report window open on your computer, you can reopen any transaction by double-clicking it. When you do that, you'll get a window showing the original transaction. You can make any corrections you need in that window.

Now let's look at each one of these reports.

Transaction Detail by Account — The Transaction Detail by Account report lists each account in your Chart of Accounts and all transactions on each account. You can use this report to find any transactions that have been posted to the wrong account. For example, all checks written to the phone company should be posted to the Telephone account. If a check for office supplies shows up under the Telephone account, you'll need to correct the transaction. Here's how to get the Transaction Detail by Account report:

▌ Click **Report Center**, or from the **Reports** menu, choose **Accountant & Taxes**, and then **Transaction Detail by Account**. Select the full month you're closing. In the examples we show here, we've used the month of December.

Figure 18-3

The Display tab window lets you specify which dates and columns will be on the Transaction Detail by Account report.

In the Display tab of the Modify Report window, under the Columns box, select only **(left margin)**, **Type**, **Date**, **Num**, **Source Name**, **Memo**, **Class** (optional), **Amount**, and **Balance** as shown in Figure 18-3. You'll have to scroll down to the bottom of the list to find Source Name, Memo, Class, Amount, and Balance.

In **Dates**, select **Custom**. In **From** and **To**, enter the month you're closing. Ignore the remaining choices.

Click **OK**.

Figure 18-4 shows an example of this customized Transaction Detail by Account report. Check the transactions on your own report and correct any that are in error.

A/R Aging Detail Report — The A/R Aging Detail report lists all unpaid customer invoices as of the last day of the month you're closing. You can use it to show any invoices that have been posted twice, missing invoices, and payments that haven't been applied to an invoice. You'll know there's a mistake if you see a payment with a negative amount. To fix this problem, double-click the negative payment amount, and at the bottom of the Customer Payment window, select the invoice this payment should be applied to.

Here's how to get an A/R Aging Detail report like the example we show in Figure 18-5:

Click **Report Center**, or from the **Reports** menu, choose **Customers & Receivables**, and then **A/R Aging Detail**.

Figure 18-4

A sample Transaction Detail by Account report you can create in QuickBooks Pro.

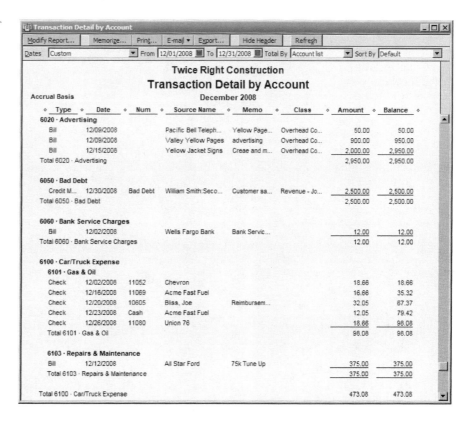

Figure 18-5

A sample A/R Aging Detail report you can create in QuickBooks Pro.

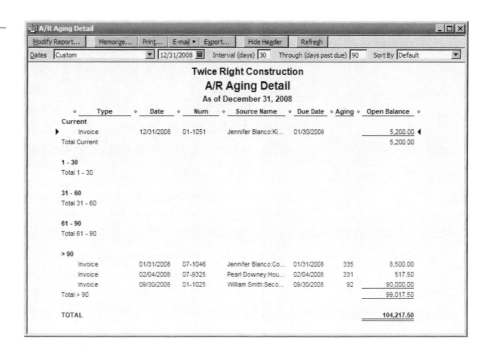

Figure 18-6

The Display tab window lets you specify which dates and columns will be on the A/R Aging Detail report.

In the Display tab of the Modify Report window, under the Columns box, select only **(left margin)**, **Type**, **Date**, **Num**, **Source Name**, **Due Date**, **Aging**, and **Open Balance** as shown in Figure 18-6. You'll have to scroll down to the bottom of the list to find Due Date, Aging and Open Balance.

In **Dates**, select **Custom**. In **From** and **To**, enter the month you're closing. Ignore the remaining choices.

Click **OK**.

Check the transactions on your own report and correct any errors.

A/P Aging Detail Report — The A/P Aging Detail report lists all unpaid vendor bills as of the last day of the month you're closing. You can use it to show any missing bills or bills that have been posted twice.

Here's how to get an A/P Aging Detail report like the one we show in our example in Figure 18-7:

Click **Report Center**, or from the **Reports** menu, choose **Vendors & Payables**, and then **A/P Aging Detail**.

In the Display tab of the Modify Report window, under the Columns box, select only **(left margin)**, **Type**, **Date**, **Num**, **Name**, **Due Date**, **Aging**, and **Open Balance** as shown in Figure 18-8. You'll have to scroll down to the bottom of the list to find Due Date, Aging, and Open Balance.

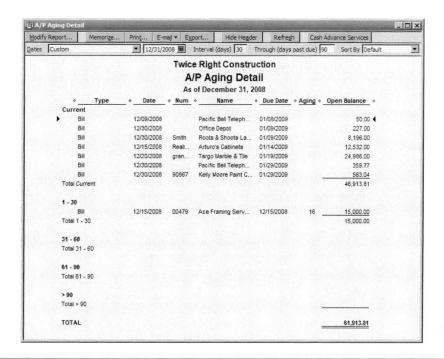

Figure 18-7
A sample A/P Aging Detail report you can create in QuickBooks Pro.

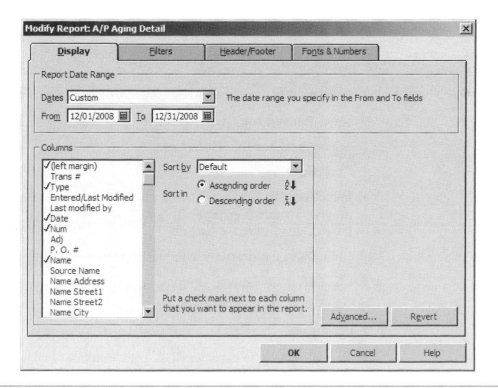

Figure 18-8
The Display tab window lets you specify which dates and columns will be on the A/P Aging Detail report.

▌ In **Dates**, select **Custom**. In **From** and **To**, enter the month you're closing. Ignore the remaining choices.

▌ Click **OK**.

Check all the transactions on your own report and correct as necessary. Be sure to watch out for missing bills and bills that have been posted twice.

Printing and Filing Monthly Reports

Now you're ready to print these reports for the month you're closing. To print the Transaction Detail by Account and the A/R and A/P Aging Detail reports, just follow the same steps you used to open and correct the reports, and then click Print in each report window.

You'll also want to print financial statements for the month. To do that:

▌ Click **Report Center**, or from the **Reports** menu, choose **Company & Financial**, and then **Balance Sheet Standard**.

▌ Click **OK** to close the Modify Report window, then click **Print**.

▌ Click **Report Center**, or from the **Reports** menu, choose **Company & Financial**, and then **Profit & Loss YTD Comparison**.

▌ Click **OK** to close the Modify Report window, then click **Print**.

Back Up Data and Close the Month

When you've made sure everything is correct, you're ready to back up your data. To do that:

▌ From the **File** menu, choose **Save Copy or Backup**.

▌ Select **Backup copy**. Click **Next**.

▌ Select **Local backup**. Click **Next**.

▌ Click **Options**.

▌ In the Backup: Options window, click **Browse** and select a location for your backup file. Typically, you will select a CD-RW or USB flash drive to store the backup.

▌ Click **OK** in the Browse for Folder window.

▌ Click **OK** in the Backup Options window.

▌ Select **Save it now**. Click **Next**.

▌ Click **Save.**

Now you're ready to close and lock yourself out of making changes to the accounting period you just completed.

▌ From the **Edit** menu, choose **Preferences**.

▌ Click the **Accounting Preferences** icon in the left hand scroll bar.

▌ Click the **Company Preferences** tab.

▌ In the Closing Date area, click on **Set Date/Password** and enter the last day of the month just closed. This will cause a warning to display if you're making any entries into the closed month. You may also want to set a password to prevent any unauthorized change of transactions in closed months.

▌ Click **OK**.

Reading and Understanding Your Financial Reports

It's important to take some time to analyze how well your business is doing before you file your reports. The rest of the work up to this point was just that, work. Now we get to have a little fun. Let's analyze the past to change the future. You can pinpoint your overexpenditures and figure out, finally, what your overhead percentage is running.

Balance Sheet

The Balance Sheet lists:

▌ your company's assets (things your company owns).

▌ liabilities (things your company owes money for).

▌ capital or equity (assets minus the liabilities or the amount of equity you've built up in the business).

There are two things to look for in this report. First, make sure everything is listed on the report. Are your checking account, savings account, company vehicles, large equipment, and so on, listed as assets? Then check the liabilities. Are all loans, lines of credit, and payroll taxes you owe listed?

Second, let's analyze the report. Keep in mind that this is the report the bank will look at to determine how well you manage your business, as well as its stability and growth. The bank will be looking closely at the assets in relation to the liabilities. The bank wants to know if your company is solvent. In other words, if you stopped business today, could your company sell its assets and pay off all its liabilities and have a few bucks left over? The bank wants to see total assets of two times your total liabilities, give or take a few dollars.

Balance Sheet Comparison (Optional Report for Analysis)

It's best if you analyze the Balance Sheet over two periods — for example first quarter 2008 and second quarter 2008 (or better yet, year to year).

If you've been using QuickBooks Pro for more than one quarter, print the Balance Sheet Comparison report to see how you did this quarter compared to last quarter. To print this report:

▌ Click **Report Center**, or from the **Reports** menu, choose **Company & Financial**, and then **Balance Sheet Prev Year Comparison**.

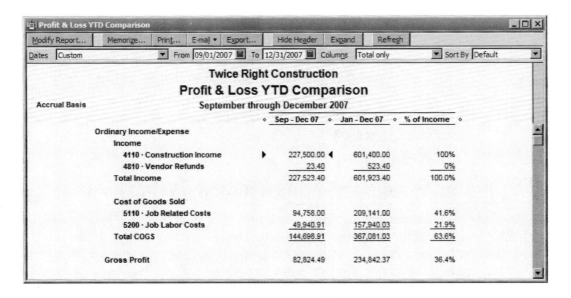

Figure 18-9
A sample Profit & Loss YTD Comparison report for the last quarter of 2007.

▌ Click **OK** to close the Modify Report window, then click **Print**.

On this report, check to see:

▌ Are the assets increasing?

▌ Are the liabilities decreasing?

▌ If the liabilities increased, the assets should have increased as well.

▌ Is the value of the business (assets minus liabilities) increasing or decreasing?

If it's decreasing, something is wrong. Give the problem your immediate attention.

Profit & Loss YTD Comparison Report

The Profit & Loss YTD Comparison report shows how much income your company has generated versus its expenses for the same time period. It's obvious here that you want to be generating more income than expenses, and the bottom line or net income/loss needs to be positive. This report will also list your total job costs and your overhead percentage. In Figure 18-9, we show an example of the report. We've modified the report to include the % of Income column.

Take a look at the Total Cost of Goods Sold (COGS) row and check the percentage in the column called % of Income. In the construction business, this percentage averages about 40 to 60 percent. If your percentage is lower,

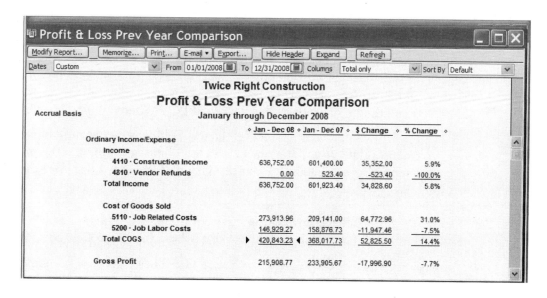

Figure 18-10

A sample Profit & Loss Previous Year Comparison report comparing 2007 with 2008.

good for you! That means more money is going in your pocket. You're spending less on construction-related costs than the average construction company.

Now scroll down to the Total Expense row and check the percentage in the column called % of Income. This is the percentage of your income you spend for overhead costs — costs that you can't associate with a job expense, such as office rent, telephone, postage, and utilities. If your percentage is 10, it means that for every $100.00 you generate in revenue, you spend $10.00 for overhead costs. In the construction business, the average is about 8 to 12 percent.

To print this report:

▌ Click **Report Center**, or from the **Reports** menu, choose **Company & Financial**, and then **Profit & Loss YTD Comparison**.

▌ Click **% of Income** box.

▌ Click **OK** to close the Modify Report window, then click **Print**.

Profit & Loss Prev Year Comparison Report

Again, like the Balance Sheet, it's best if you analyze the Profit & Loss over two years. If you've been using QuickBooks Pro for more than one year, print the Profit & Loss Prev Year Comparison report to see how you did this year compared to last year. In Figure 18-10, we show an example of a report comparing 2007 with 2008.

This report will tell you how well you did this year compared to last year. First, compare your Total Income. Has it increased or decreased from last year? Second, compare your Total COGS — where COGS = Job Costs. Have your job-related costs gone up or down? Why would that be? Third, look at Total Expense. Have your expenses gone up or down? Are you spending more money this year than last? Where are you spending less?

File the monthly reports in the company accounting binder for the appropriate year. Label the tabs in the binder according to the names of the reports you printed — Balance Sheet and Profit & Loss, for example.

To print this report:

▌ Click **Report Center**, or from the **Reports** menu, choose **Company & Financial**, and then **Profit & Loss Prev Year Comparison**.

▌ Click **OK** to close the Modify Report window, then click **Print**.

End of Year Procedures

In QuickBooks Pro you don't have to end the current year before you start a new year. To end your year:

▌ Follow the procedures in the previous section to close each month.

▌ Write off bad debts.

▌ Print your financial reports, file them, and give a copy to your tax preparer.

▌ Back up your QuickBooks file and close and lock the year.

Write Off Bad Debts and Run the Aging Detail Reports

At the end of the year you'll want to write off any bad debts from customers you know aren't going to pay you. To do this, create a credit memo for each customer invoice that won't be paid. This will allow you to clear the invoice out of Accounts Receivable and decrease your net profit by the amount of the unpaid invoice.

A/R Aging Detail Report

▌ Run the A/R Aging Detail report, making sure the **To** and **From** dates are both the last day of the fiscal year. Make sure every invoice listed was actually outstanding as of the last day of your fiscal year. Check that you've entered all customer payments.

Write Off Bad Debts

Write off any bad debts from customers you know aren't going to pay. To do this, create a credit memo for each customer invoice that won't be paid.

Figure 18-11

Here's an example of how to set up a Bad Debt account.

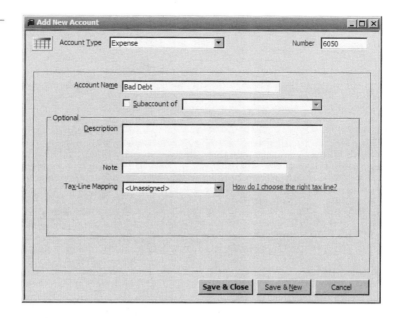

This will allow you to clear the invoice out of Accounts Receivable and decrease your net profit by the amount of the unpaid invoice. To begin, create an expense account for the bad debt:

▌ Click **Home**, then click the **Chart of Accounts** icon, or from the **Lists** menu, choose **Chart of Accounts**.

▌ Pull down the **Account** menu and choose **New**.

▌ Select the **Expense** category. Click **Continue**.

▌ Fill in the Add New Account window as shown in Figure 18-11. We've used 6050 as our Bad Debt account.

▌ Click **Save & Close**.

Next, create an Item for bad debt:

▌ Click **Home**, then click the **Items & Services** icon, or from the **Lists** menu, choose **Item List**. In the Item List window, pull down the **Item** menu and choose **New**.

▌ Fill in the New Item window as shown in Figure 18-12. Link the item to the Bad Debt expense account you created above.

▌ Click **OK**.

Next, create a credit memo to offset the amount the customer will not be paying:

Figure 18-12

Here's an example of how to create a Bad Debt item linked to a Bad Debt account.

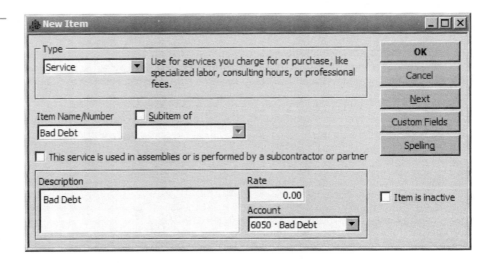

■ Click **Home**, then click the **Refunds & Credits** icon, or from the **Customers** menu, choose **Create Credit Memo/Refunds**.

■ Enter the **Customer:Job**.

■ Enter the date as the last day of the year or the last day of your fiscal year.

■ In the **Credit No.** field, enter **Bad Debt**.

■ Under **Item**, select **Bad Debt** from the pull-down Item list.

■ In **Amount** enter the amount the customer was invoiced but will not be paying you. See Figure 18-13.

■ Click **Save & Close**.

Now you'll be prompted with an Available Credit dialog box. See Figure 18-14.

■ Select **Apply to an invoice**.

■ Click **OK**.

Finally, in the Apply Credit to Invoices window:

■ Put a checkmark in the left hand column to apply the credit to the appropriate invoice(s). See Figure 18-15.

■ Click **Done**.

A/P Aging Detail Report

■ Run the A/P Aging Detail report, making sure the **To** and **From** dates are both the last day of the fiscal year. If there are duplicates or mistakes, fix them before you close the year.

Figure 18-13

Use a credit memo to record the bad debt.

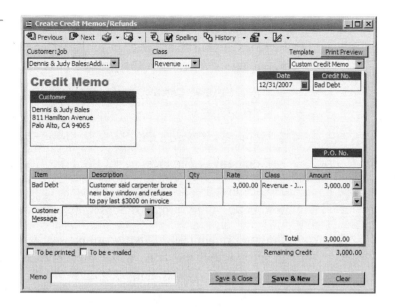

Figure 18-14

The Available Credit dialog box.

Figure 18-15

Apply the credit to the appropriate invoice(s).

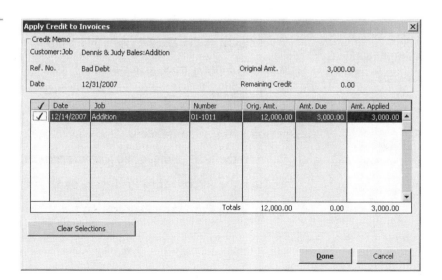

Print Your Annual Reports

At the close of your fiscal year, print the Balance Sheet, Profit & Loss statement, and aging reports for the year you're closing.

▌ Click **Report Center**, or from the **Reports** menu, choose and print these reports:

▌ Company & Financial, Balance Sheet Standard

▌ Company & Financial, Profit & Loss Standard

▌ Customers & Receivables, A/R Aging Detail

▌ Vendors & Payables, A/P Aging Detail

File the annual reports in the company accounting binder for the appropriate year. Label the tabs in the binder according to the names of the reports you printed — Balance Sheet, Profit & Loss, for example.

Make a copy of the Balance Sheet and Profit & Loss reports for your tax preparer.

Back Up Your Data Files and Close the Year

To back up your end of the year data:

▌ From the **File** menu, choose **Save Copy or Backup**.

▌ Select **Backup copy**. Click **Next**.

▌ Select **Local backup**. Click **Next**.

▌ Click **Options**.

▌ In the Backup: Options window, click **Browse** and select a location for your backup file. Typically, you will select a CD-RW or USB flash drive to store the backup.

▌ Click **OK** in the Browse for Folder window.

▌ Click **OK** in the Backup Options window.

▌ Select **Save it now**. Click **Next**.

▌ Click **Save**.

Now you're ready to close and lock yourself out of making changes to the period you just completed.

▌ From the **Edit** menu, choose **Preferences**.

▌ Click the **Accounting Preferences** icon in the left hand scroll bar.

▌ Click the **Company Preference** tab.

▌ In the Closing Date area, click on **Set Date/Password** and enter the last day of the year just closed. This will cause a warning to display if you're making any entries into the closed month. You may also want to set a password to prevent any unauthorized change of transactions in closed months.

▌ Click **OK**.

In this chapter, you learned what we recommend you do at month end and at year end. You saw how to reconcile your bank statements, print financial reports, and review transactions to make sure they were entered correctly. The next chapter is for builders who develop land. If you're a builder, be sure to go through this chapter. If you're a remodeling contractor and you don't develop land, nor plan to, you may want to just skim over the next chapter. But read the conclusion. It's pertinent to all contractors.

Real Estate Development

As a spec builder or developer, you have to make sure your accounting system handles some specialized tasks. You need to track the money you borrow and spend to purchase land, improve that land, and then build on the improved land. Then you also need to record the sale of the property. In this chapter, we'll look at how you can use QuickBooks Pro to easily deal with these matters.

New Accounts

First, you may need to add a few new accounts to your Chart of Accounts. If you need more information on setting up accounts, see Chapter 3. Here are the new accounts you'll need:

Account number	Name	Type
1111	Adjustment Register	Bank
1460	Escrow Deposit	Other Current Asset
1470	Land Purchase	Other Current Asset
1471	Land Interest/Closing Costs	Other Current Asset
1480	WIP — Land Development	Other Current Asset
1490	WIP — Construction	Other Current Asset
2300	Loans Payable	Other Current Liability
2400	Land Acquisition Loan	Other Current Liability
2405	Land Development Loan	Other Current Liability
2410	Construction Loan	Other Current Liability

▍ *Adjustment Register* — Use the Adjustment Register account to temporarily record the transaction where you pay off one loan with another loan. For example, a development loan may pay off a land loan, or a construction loan may pay off a development loan. You'll see later exactly how to use this account to temporarily record loans that have been paid off during escrow closing. In our example Chart of Accounts, we use 1111 Adjustment Register.

■ *Escrow Deposit Account* — Use the Escrow Deposit account if you're required to place a deposit with an escrow company when you purchase land. Usually you write a check to the escrow company to open up an escrow account. In our example Chart of Accounts, we use 1460 Escrow Deposit.

■ *Land Purchase Accounts* — Use the Land Purchase accounts to keep track of the original purchase price of any land you purchase for development and costs related to that purchase. In our example Chart of Accounts, we use 1470 Land Purchase and 1471 Land Interest/Closing Costs.

■ *WIP — Work in Process Accounts* — Use the WIP accounts to keep track of the costs of developing and building on a property. These accounts will appear on your balance sheet as assets, even though you use them for costs. In real estate development, you have to hold your costs until you sell the property. Then you match the sale of the building and the construction costs in the same time period. Once you sell the building, you'll transfer the WIP totals from the WIP accounts to the Job Related Costs account. In our example Chart of Accounts, we use 1480 WIP Land Development and 1490 WIP Construction.

■ *Loan Accounts* — The loan accounts should total the amount of funds you borrowed for land acquisition, land development, and building construction. These loan accounts appear on your balance sheet as liabilities. When you pay off a loan or sell a property, the loan balances for the property will return to 0. We'll explain this rather complicated process, step-by-step, later in this chapter in the section on Recording the Sale of a Property. In our example Chart of Accounts, we use 2400 Land Acquisition Loan, 2405 Land Development Loan, and 2410 Construction Loan.

Setting Up a Development Job

To track the costs and income for each individual building in a development, set up each one as a customer. For example, our Twice Right Construction Company is developing property on Ellis Street. Since we don't know who the customer is yet, we set up a customer called Ellis Street. If you have more than one job in a single development, you can make the development the customer and set up the properties as individual jobs. For example, if Twice Right purchased three lots on Ellis Street, the customer would be Ellis Street Development and each job would be its street address — 1120 Ellis Street, 1130 Ellis Street, and 1140 Ellis Street, as shown in Figure 19-1. However, in most cases you'll be using a lot and block number, i.e., Lot 1, Block 10 or L1B10.

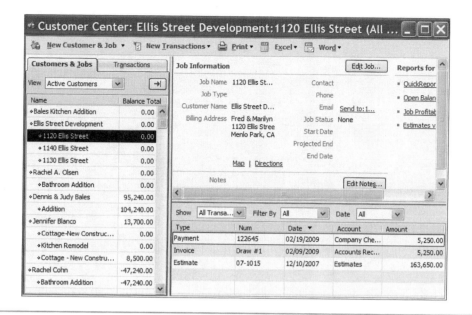

Figure 19-1
Notice that Ellis Street Development is the customer and the jobs are listed as the address of each piece of property. If you don't have an address for the property, you may have to list each job as Lot 1, Lot 2, etc.

Using Items to Track Construction Costs as WIP

The methods we show you here are just for a spec builder or developer who is setting up and linking items (job phases of construction). This is different from the way we showed you to link items in Chapter 4. In this section, we'll use WIP and Construction Loan accounts.

If you want to track the costs of development by phase you need to create items and subitems for each phase. Now let's see how to set up a new item for 34 Land Dev (short for Land Development Costs), and a subitem 34.1 Loan Fee (short for Land Development Loan Fees) as shown in Figures 19-2 and 19-3. To create the new item:

▮ Click **Home**, then click **Items & Services**, or from the **Lists** menu, choose **Item List**. In the **Item List** window, pull down the **Item** menu and choose **New**.

▮ In **Type**, select **Service**.

▮ In **Account**, enter a Land Purchase, Land Interest/Closing Costs, WIP — Land Development, or WIP — Construction account. In our example, we selected 1480 WIP — Land Development because all the costs related to this item should get classified as land development costs.

▮ Fill in the other necessary fields and click **OK**.

Figure 19-2

When you set up an item, be sure to link it to the correct asset account.

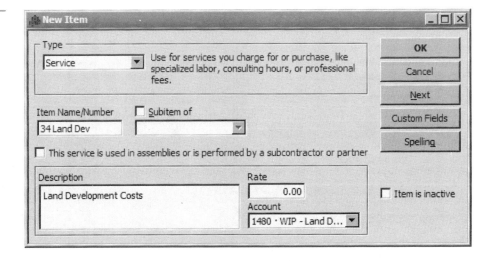

Figure 19-3

Here's an example of a new item that's a subitem. Notice that the "Subitem of" box is checked.

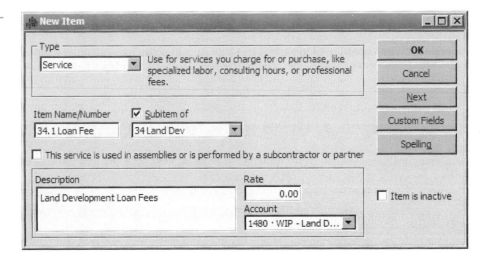

Keep in mind that each item you set up as a spec builder needs to be set up with an asset account (in the Account box). Link land purchase items to a land purchase asset account such as 1470 Land Purchase. Link land development items to a WIP development asset account such as 1480 WIP Development. Link construction cost items to a WIP construction asset account such as 1490 WIP Construction and a construction liability loan account such as 2410 Construction Loan.

Use the same procedure to create subitems, remembering that you have to set up an item before you can set up any subitems under it. Figure 19-3 shows how we created 34.1 Loan Fee as a subitem of the 34 Land Dev item. Later in this chapter, we'll add a few more items that we'll use when the property is sold.

Land Purchase Transactions

Earnest Money

When you write a check to an escrow company to open escrow to purchase a piece of property, the money is called earnest money. In effect, it's a good faith deposit. Here's how to record the transaction and write an earnest money check:

▌ Click **Home**, then click the **Write Checks** icon, or from the **Banking** menu, choose **Write Checks**. See our example in Figure 19-4.

▌ At the top of the Write Checks window, in **Bank Account**, make sure you select the checking account you write the check from.

▌ In **Pay to the Order of**, enter the name of the escrow company. Enter the check amount beside the company name.

▌ Make sure the check number is accurate or leave it blank if you're going to print the check later.

▌ In **Memo**, enter a description, such as land address, and a brief statement saying the check is for escrow earnest money.

▌ Click the **Expenses** tab.

▌ Click in the **Account** column and select your Escrow Deposit account from the account list. In our example, we used 1460 Escrow Deposit.

▌ In **Amount**, enter the full amount of the check.

▌ Leave Customer:Job and Class blank.

▌ Click **Save & Close**.

Figure 19-4

Here's an example of how to record earnest money you pay to a title company to start escrow.

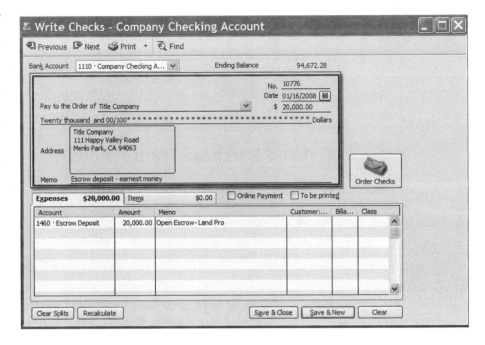

Closing Escrow on a Land Purchase

You'll need to write another check to the escrow company when you complete the land purchase and the title to the land is switched into your company's name. When you write this check, you can also record that you purchased the land and paid closing costs. And if you have a seller carry back loan or a loan from a bank, you can also record that on the check to the escrow company. Let's work through our example so you can see how to do this. In Figure 19-5, we show you an escrow closing sheet you can use to follow our example. Here are the steps:

▌ Click **Home**, then click **Write Checks**, or from the **Banking** menu, choose **Write Checks**. See our example in Figure 19-6.

▌ In the Write Checks window, in **Bank Account**, select the account that you're issuing the funds from.

▌ In **Pay to the Order of**, enter the name of the escrow company.

▌ In the **$** field, enter the amount of the check you write to the escrow company when the escrow on the land closes.

▌ Click the **Items** tab.

▌ Click the **Item** column and select your land purchase item from the pull-down list.

Figure 19-5

Here's an example of an escrow closing statement for a land purchase.

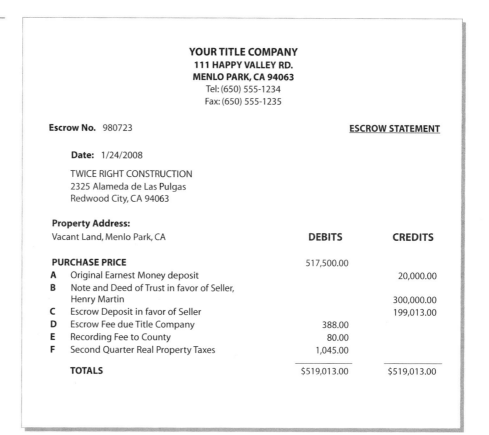

YOUR TITLE COMPANY
111 HAPPY VALLEY RD.
MENLO PARK, CA 94063
Tel: (650) 555-1234
Fax: (650) 555-1235

Escrow No. 980723   **ESCROW STATEMENT**

Date: 1/24/2008

TWICE RIGHT CONSTRUCTION
2325 Alameda de Las Pulgas
Redwood City, CA 94063

Property Address:

		DEBITS	CREDITS
Vacant Land, Menlo Park, CA			
PURCHASE PRICE		517,500.00	
A	Original Earnest Money deposit		20,000.00
B	Note and Deed of Trust in favor of Seller, Henry Martin		300,000.00
C	Escrow Deposit in favor of Seller		199,013.00
D	Escrow Fee due Title Company	388.00	
E	Recording Fee to County	80.00	
F	Second Quarter Real Property Taxes	1,045.00	
	TOTALS	$519,013.00	$519,013.00

Figure 19-6

Use the Items tab to record land purchase and closing costs. In this example, we used items 31.1 Land Purchase Price and 31.2 Land Closing Costs. Your items for land purchase and closing costs may be different.

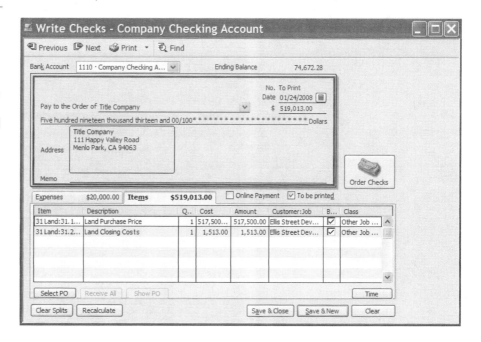

▌ In **Amount**, enter the purchase price from the escrow settlement sheet. In Figure 19-5, it's $517,500.

▌ Click the **Customer:Job** column and select the customer for the transaction from the pull-down list.

▌ Click the next line in the **Item** column and select your land closing costs item from the pull-down list.

▌ In **Amount**, enter the sum of closing fees (lines D - F in Figure 19-5).

▌ Click the **Customer:Job** column and select the customer for the transaction from the pull-down list.

If the seller carries back a loan, you record the loan amount on the Expenses tab. Here's how you can do that using the Expenses tab on the check shown in Figure 19-6:

▌ Click the **Expenses** tab. See Figure 19-7.

▌ Click in the **Account** column and select your Escrow Deposit account from the pull-down list. It's 1460 in our example.

▌ In **Amount**, enter the escrow earnest money as a negative number. In our example, it's −$20,000 (line A in Figure 19-5).

▌ Click the next line in the **Account** column and select the Land Acquisition Loan account from the pull-down list. It's 2400 in our example.

▌ In **Amount**, enter the total amount of the loan or seller carry back amount, as a negative number. In our example, it's −$300,000 (line B in Figure 19-5).

▌ Leave **Class** blank.

▌ Click **Save & Close**.

After you click Save & Close, you'll have recorded that you closed escrow on the land, recorded a seller carry back loan (if needed), recorded the closing costs, and lowered the checking account balance by the amount of the check you wrote to the escrow company.

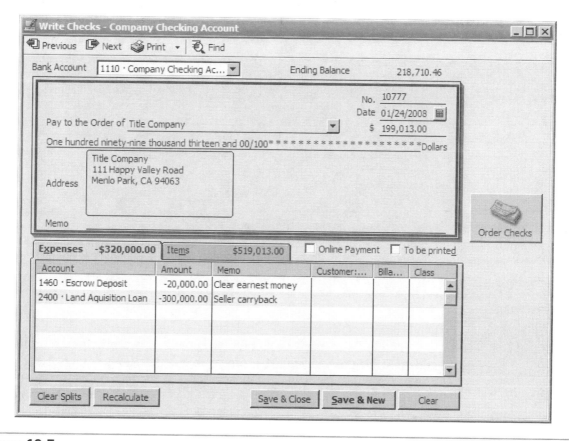

Figure 19-7

Here's how to enter a seller carry back loan on land and how to clear out the earnest money involved in the sale.

Personal Loans

Most real estate developers or spec home builders need to borrow money to buy land or do some developing before they take out a construction or development loan. These are basically personal loans. Whenever you get a personal loan (not including a construction loan) from a new entity, you need to add a new liability account to your Chart of Accounts. For example, let's say that our Twice Right Construction Company borrowed $25,000 from an investor. The investor is Paul Olsen, a long-time friend. We need to keep track of this money after we deposit it in our bank account. So we need to set up a new liability account for the loan.

However, if you don't have a loans payable header account in your Chart of Accounts, you should set that up first. In our example, we use 2300 Loans Payable as the loans payable header account. Then, we use the sub-account 2320 for Paul Olsen's personal loan to the business.

Figure 19-8

Create a liability account for each personal loan from a new entity.

Now, on to the subaccount for the entity you borrowed money from. Keep in mind that this loan could be from an individual, another company, or a bank. In our example, the loan is from an individual, but the procedure would be the same if it were from another entity. However, don't set up a new account if the loan is a construction loan. Construction loans are handled differently and will be covered later in this chapter, under the heading Construction Loans.

▌ Click **Home**, then click the **Chart of Accounts** icon, or from the **Lists** menu, choose **Chart of Accounts**.

▌ In the Chart of Accounts window, pull down the **Account** menu and choose **New**.

▌ From the **Other Account Types** pull-down list, select **Other Current Liability**. Click **Continue**.

▌ In **Number**, enter the new account number. In our example it's 2320.

▌ In **Name**, enter Loan – name of entity (Paul Olsen in our example). See Figure 19-8.

▌ Click the **Subaccount of** box and select the loans payable header account number from the pull-down list. It's 2300 in our example.

▌ Click **Save & Close**.

Follow the same procedure for any other entities that loaned you money. Keep in mind that you only enter new accounts for loans that are not construction or development loans. Construction, development, and seller carry back loans on land purchases are usually handled differently from personal loans. Construction loans, for example, have draw schedules and have to be paid back during the close of escrow. The personal loans we set up here are typically more informal than construction loans.

Figure 19-9

Here's how to record the deposit of a personal loan.

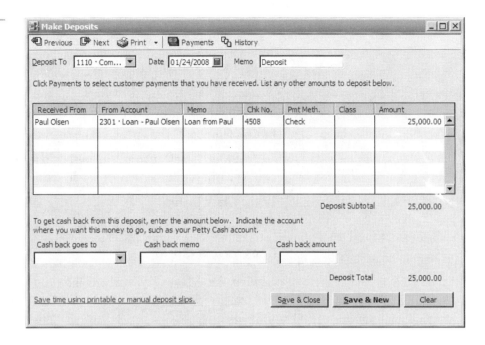

Depositing a Personal Loan

To record that you've deposited the money you received from a personal loan, you'd follow the instructions below. In our example, we deposit the $25,000 that Paul Olsen loaned us.

First, make sure that this deposit is separate from your ordinary income deposits. For example, if you received three checks that need to be deposited, two checks from customers and one personal loan check, make sure you use a separate deposit slip for the personal loan check. The other two customer checks can be added together on the same deposit slip, but the personal loan has to be deposited separately. This is because it will be recorded individually in the QuickBooks Pro checkbook reconciliation, and it would be best if the bank recorded it the same way.

▌ Click **Home**, then click the **Record Deposits** icon, or from the **Banking** menu, choose **Make Deposits**.

▌ In the Make Deposits window, from the pull-down **Deposit To** list, make sure you choose the checking account that you're going to deposit the money to. In our example shown in Figure 19-9, we use 1110 Company Checking.

▌ In **Date**, enter the date you'll be taking the deposit to the bank.

- In **Received From**, enter the entity you got the money from (not the job where you'll be using the money).

- In **From Account**, select the loan payable account from the pull-down list for the entity. In our example, it's 2320 Loans - Paul Olsen.

- In **Memo**, enter what you want to see in your checkbook register.

- In **Chk No.**, enter the check number from the entity.

- In **Pmt Meth.**, select from the pull-down list whether you received a check, cash, money order, or credit card payment.

- Leave **Class** blank.

- In **Amount**, enter the amount you're depositing.

- Click **Save & Close**.

Following these steps will record that you received and deposited the personal loan, and that you owe that entity for the personal loan. Whenever you print a Balance Sheet, the liability for this personal loan will appear, reminding you that you have to pay this personal loan back at some point.

Recording an Interest Payment on a Personal Loan

You may be required to pay interest on a personal loan on a periodic basis such as monthly, quarterly, or yearly. Or you may not have to pay both principal and interest until you pay back the personal loan. Regardless of when you pay the interest, the procedure is the same. You enter one transaction for the interest payment and one for the principal payment. However, the check you give to the entity may include just interest, or both interest and principal.

To record that you owe an interest payment on a personal loan (even if you may not be writing the check for some time) you'd enter a bill following the steps below. The bill you enter will track the interest expense, a job expense on the job reports (if a job is associated with the expense), and a transaction in your Accounts Payable (for interest only) to the entity you borrowed the money from.

- Click **Home**, then click **Enter Bills**, or from the **Vendors** menu, choose **Enter Bills**.

- In the Enter Bills window, from the pull-down Vendor list, select the entity you owe the interest payment to. In our example shown in Figure 19-10, it's Paul Olsen.

- In **Date**, enter the date you owe the interest.

- If the entity has given you a bill or invoice for the interest, enter that number in the **Ref. No** field. In our example, we didn't have a bill from Paul Olsen, so we entered Interest.

- In **Amount Due**, enter the amount of the interest only.

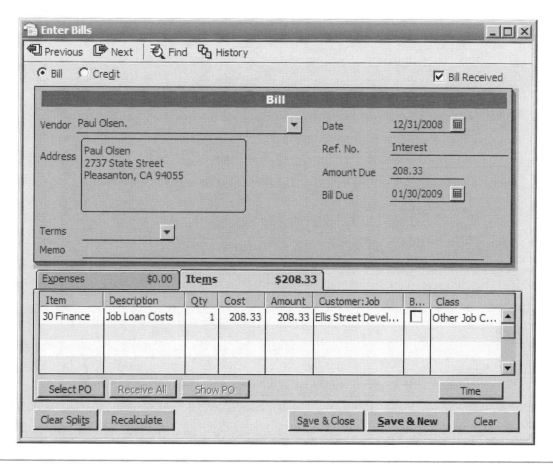

Figure 19-10
Here's how to record interest due on a loan.

▌ In **Memo**, enter a description for the bill.

▌ Click the **Items** tab.

▌ Click in the **Item** column and from the pull-down **Item** list, select the appropriate item.

▌ Enter **Amount**, **Customer:Job**, and **Class**. This will capture the interest cost on job cost reports.

▌ Click **Save & Close**.

Entering this bill will record that you owe the entity for the interest. The bill will show up in the Pay Bills window when it's due.

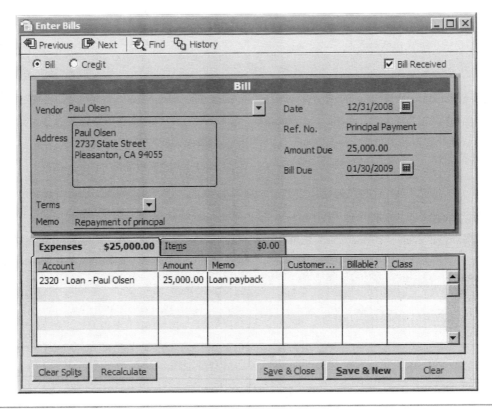

Figure 19-11
Here's how to record paying back a personal loan.

Recording a Principal Payment on a Personal Loan

When it comes time to pay down (or pay off) the personal loan, you'll need to follow these steps:

▌ Click **Home**, then click **Enter Bills**, or from the **Vendors** menu, choose **Enter Bills**.

▌ In the Enter Bills window, from the pull-down **Vendor** list, select the entity you'll be paying the principal payment to. In our example shown in Figure 19-11, it's Paul Olsen.

▌ In **Date**, enter the date you decide to pay the principal.

▌ In **Ref. No**, enter the loan number or a description of the personal loan. In our example we didn't have a loan number, so we entered Principal Payment.

▌ In **Amount Due**, enter the amount of the principal only.

▌ In **Memo**, enter a description for the bill.

▌ Click the **Expenses** tab.

▌ Click in the **Account** column and select the account you set up when you borrowed the money. In our example, the account is 2320 Loan - Paul Olsen.

▌ Click **Save & Close**.

Entering this bill will record that you owe that entity for a principal payment. The bill will show up in the Pay Bills window when it's due.

Development Loans

A development loan is a loan you take out specifically to develop the infrastructure of a land parcel, such as subdividing the land into lots, sidewalks, roads, common areas, bridges, parks, etc. Usually the lender of a development loan requires you to also pay off the land loan as part of the development loan. So you also have to include the balance of the land loan in the amount you borrow for a development loan.

Closing Escrow on a Development Loan

There are several transactions that you need to record when you close escrow on a development loan. They are:

1. Record loan fees, including loan origination, document prep, processing, and underwriting fees.

2. Record escrow fees.

3. Record title insurance fees.

4. Record miscellaneous fees.

5. Record property tax payment.

6. Record paying off the seller carry back loan on the land. Skip this step if there's no loan on the land.

7. Record interest on the land loan. Skip this step if you don't have an existing land loan.

We'll take you through an example to show you how to do all this. You can use the Escrow Statement shown in Figure 19-12 to follow our example. To begin the process, we'll write the check shown in Figure 19-13.

▌ Click **Home**, then click **Write Checks**, or from the **Banking** menu, choose **Write Checks**.

▌ In the Write Checks window, from the pull-down **Bank Account** list, select your Adjustment Register account if the development loan paid off the land loan. If you didn't pay off the land loan with the development loan and you wrote a check at the close of escrow, select the account you wrote that check from.

YOUR TITLE COMPANY
111 HAPPY VALLEY RD.
MENLO PARK, CA 94063
Tel: (650) 555-1234
Fax: (650) 555-1235

Escrow No. 123456-ABC   ESCROW STATEMENT

Date: 3/15/2008

TWICE RIGHT CONSTRUCTION
2325 Alameda de Las Pulgas
Redwood City, CA 94063

Property Address:

Vacant Land, Menlo Park, CA		**DEBITS**	**CREDITS**
YOUR ESCROW CO. Loan Proceeds			756,000.00
A	Loan Origination Fee	22,680.00	
B	Document Preparation Fee	650.00	
C	Processing Fee at 0.5%	2,340.00	
D	Fed Ex and Corp. Verif.	45.00	
E	Processing & Underwriting Fee	800.00	
F	Endorse (104.1) & Coll Set up	796.00	
G	Escrow Fee to Your Escrow Co.	250.00	
H	Title Insurance to Your Escrow Co.	1,266.50	
I	Inspection Fee	100.00	
	Record Documents	20.00	
	Reconveyance Fee	110.00	
	Recording Specific Guaranty	10.00	
J	Property Taxes paid to the County	4,914.97	
K	Seller carry back loan payoff		
	Principal	248,863.06	
L	Interest due at lot loan payoff date	5,178.20	
M	Funds Available	467,976.27	
	TOTALS	$756,000.00	$756,000.00

Figure 19-12
Here's an escrow statement you can use to follow our example.

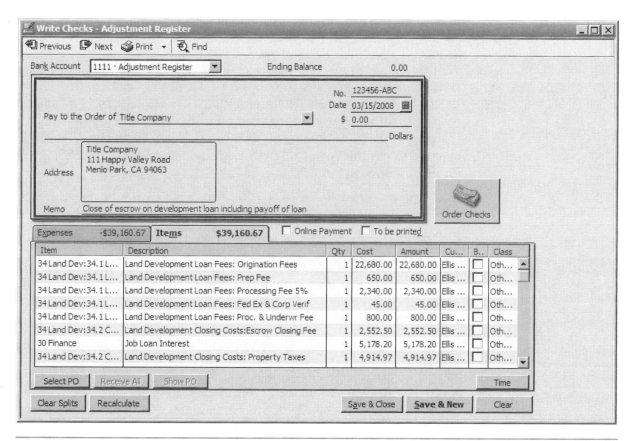

Figure 19-13
Use the Write Checks window to enter information from an escrow closing statement.

❚ In **$**, enter the amount of the check you wrote at the close of escrow. If you didn't have to write a check at the close of escrow because the new loan paid all of your costs, enter 0 as shown in the example. See Figure 19-13.

❚ In **No.**, enter the escrow number from the escrow statement.

❚ In **Date**, enter the date of the escrow statement.

❚ In **Pay to the Order of**, enter the name of your escrow company.

❚ In **Memo**, enter a description such as close of escrow.

Recording Fees and Taxes on a Development Loan

Now let's enter the fees and taxes shown on the escrow closing statement. We'll go to the Items tab and enter lines in the Item column for these amounts. All the amounts entered here (except the land loan interest on line L) are charged to a subitem of 34 Land Dev Costs item. The amount on line L is the interest on the loan up to the escrow funding date. You charge it to item 30 Finance. If you have a loan on the land, you'll be required to pay this interest. In our example, we paid the seller $5,178.20 for interest through escrow.

In our example, the property taxes hadn't been paid for the prior year, so there was a penalty tax payment due (line J). Also, the taxes needed to be prepaid for the current year.

▮ Click **Items** tab.

▮ Click in the **Item** column and enter necessary amounts in the **Amount** column. In our example shown in Figure 19-13, we enter amounts from lines A through J shown in Figure 19-12.

Recording Paying off a Land Loan

Now we're ready to record the payoff of the loan on the land. We'll do that on the Expenses tab of the check shown in Figure 19-14. Since the full purchase price of the land was entered earlier (see the section titled Closing Escrow on a Land Purchase which shows how we entered the full land value shown in Figures 19-5 and 19-6), all we want to do here is record that we paid off the land loan. In our example in Figure 19-12, the seller carry back land loan balance was $248,863.06.

▮ Click the **Expenses** tab. See Figure 19-14.

▮ Click in the **Account** column and enter your land acquisition loan account. In our example, it's 2400 Land Acquisition Loan.

▮ In **Amount**, enter the total amount of the loan payoff. In our example shown in Figure 19-14, the amount is $248,863.06.

▮ If the development loan paid these costs, you need to select account 2405 Land Development Loan and enter as a negative number the total of the fees (from the Expenses tab) and the amount the bank paid on your behalf to pay off the land loan. This will increase the loan balance.

▮ Click **Save & Close**.

The total amount of the check should be zero *or* if you wrote a check at the close of escrow to cover some costs, the check should be made out for the amount you paid at the close of escrow.

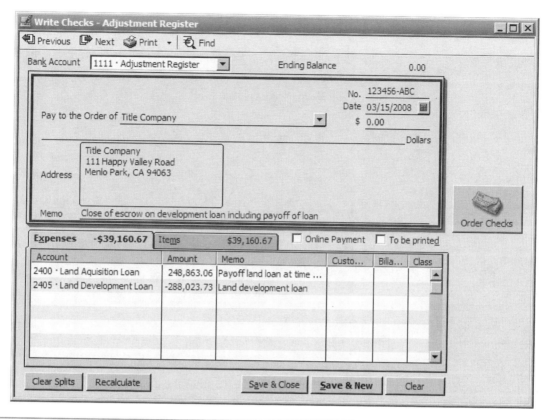

Figure 19-14

To enter an amount from an escrow closing statement use the Expenses tab of a Write Checks window.

Construction Loans

Entering a Construction Draw

When you draw funds for a development project, you need to create an invoice before you deposit the funds. And before you create the invoice, you need to create an estimate for the project. Look back to Chapter 12 if you need help with estimates. Here's how to enter an invoice for a construction loan draw:

▌ Click **Home**, then click **Invoices**, or from the **Customers** menu, choose **Create Invoices**.

▌ In the Create Invoices window, select the **Customer:Job** you're requesting the funds for. In our example, we selected Ellis Street Development — 1120 Ellis Street.

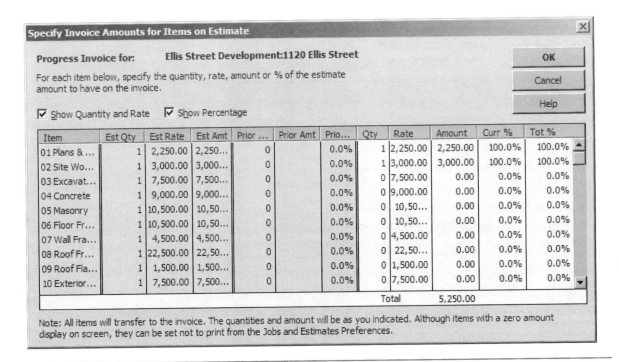

Figure 19-15
Use the Specify Invoice Amounts for Items on Estimate window to enter the amount or % for each line item of the draw.

▌ If you entered an estimate for this job, you'll see the Available Estimates dialog box. Select the estimate you would like to bill from.

▌ Click **OK**.

▌ In the Create Progress Invoice Based on Estimate window, select **Create invoice for selected items or for different percentages of each item**.

▌ Click **OK**.

▌ In the Specify Invoice Amounts for Items on Estimate window, in **Amount**, enter the amount you're billing for. In our example, shown in Figure 19-15, the draw is for $5,250 for item 01 Plans and item 02 Site Work.

▌ Click **OK**.

Back in the Create Invoices window, make sure the correct amounts appear in the draw invoice. Since each job phase item was originally linked to the Construction Loan account, when this invoice gets generated the item will automatically track the draw as a liability.

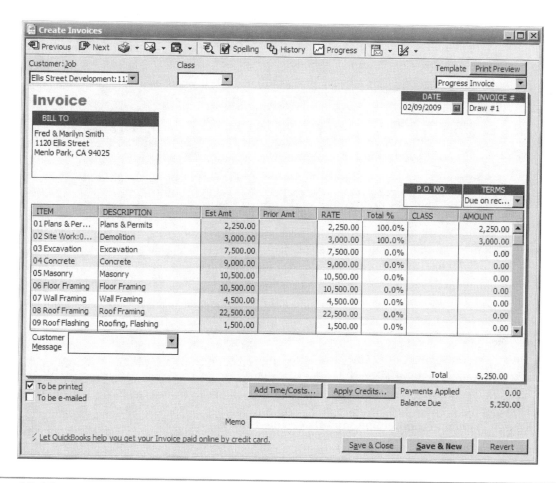

Figure 19-16
Use the Create Invoices window to record a construction draw.

▌ In **Bill To**, enter the bank or investor you're requesting the funds from. In our example, shown in Figure 19-16, we selected Fred and Marilyn Smith.

▌ Leave **Class** blank.

▌ Click **Save & Close** to create the invoice.

Depositing the Draw Check into Your Account

Here's what you need to do when you receive the draw check from the bank:

▌ Click **Home**, then click **Receive Payments**, or from the **Customers** menu, choose **Receive Payments**.

▌ From the pull-down **Received From** list, select the same customer that's on the draw request. You should see the original draw request appear, as shown in Figure 19-17.

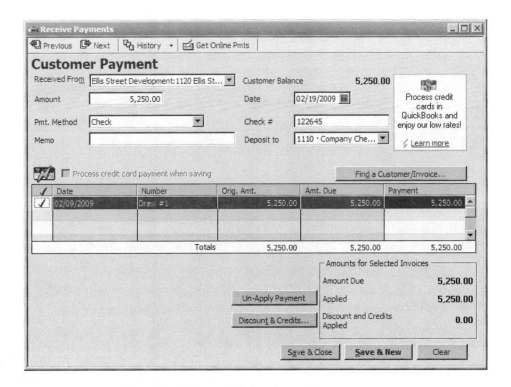

Figure 19-17
Use the Receive Payments window to record a draw request check.

▌ Fill in the date you received the payment, the amount paid, the payment method, and the check number if paid by check.

▌ In **Memo**, enter the draw request number and the bank or investor you requested funds from.

▌ Click **Save & Close**.

Recording the Sale of a Property

When you sell the property, you need to:

1. Record the sale of the property.

2. Record paying off the construction loan.

3. Account for the closing costs.

4. Transfer the WIP accounts to the Cost of Goods Sold account.

This looks complicated, but we'll show you how to do it all in just one transaction. The transaction will be on the invoice you issue for the sale of the property. Shortly, we'll walk you through the process of entering the invoice in QuickBooks Pro. First, you need to set up these seven items that you need for the process. Use the usual method to set up these items. Make sure you link the items to the proper accounts.

- *Building Sale* — link to your Construction Income account (4110 in our Chart of Accounts).

- *Building Costs* — link to your Job Related Costs account (5110 in our Chart of Accounts).

- *WIP Trans* — link to your WIP Construction account (1490 in our Chart of Accounts).

- *Closing Costs* — link to your Job Related Costs account (5110 in our Chart of Accounts).

- *Const. Draws* — link to your Construction Loan account (2410 in our Chart of Accounts).

- *Land Cost* — link to your Job Related Costs account (5110 in our Chart of Accounts).

- *Land Int* — link to your Job Related Costs account (5110 in our Chart of Accounts).

Recording an Invoice for the Sale of a Property

Now you're ready to create an invoice to record the sale and closing costs, and to clear the WIP assets and Construction Loan accounts. For our example, we created an invoice for 1120 Ellis Street when we received a check from the title company that closed the escrow on the new house. The invoice is shown in Figure 19-18. Let's take a close look at each item in the figure:

- *Building Sale* — This is the sale price of the property. Amount is the selling price of the home.

- *Land Cost* — Records the purchase price of the land in a Cost of Goods Sold (job related costs) account.

- *31.1 Land item* — This line clears the purchase price of the land out of the asset account. If you purchased several parcels in a subdivision, then the price is a portion of the original purchase price. For example, in Figures 19-5 and 19-6, we purchased land that was subdivided into five lots. Each lot was similar in size, so we divided the $517,500 by 5 to get an individual lot price of $103,500.

- *Land Int* — Records the land interest and closing costs in a Cost of Goods Sold (job related costs) account.

- *31.2 Land Int item* — This line clears out the land closing costs and interest through the end of the life of the land loan and divides the costs by the five lots. In our case, we paid $1,513 for closing costs plus $23,450 in interest. Divide that by the five lots and you have $4,992.00 per lot.

Figure 19-18

Use the Create Invoices window to record the sale of a property.

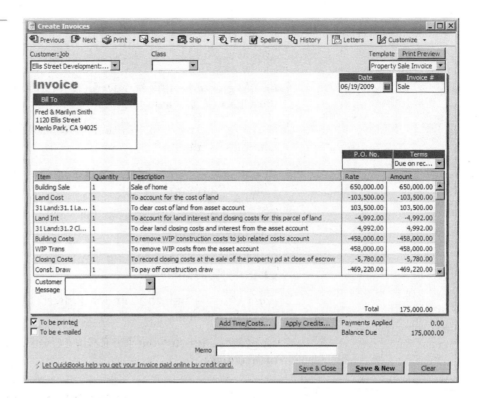

I *Building Cost* — This moves the WIP costs to the Cost of Goods Sold account. Amount is the same as the amount in the next line except it's a negative number.

I *WIP Trans* — This line clears the WIP Construction account total out of the asset account. Amount is the total WIP costs for the project. To figure out the total WIP costs for the job, print out a Profitability Detail Report. Chapter 17, Reports, has more information on this.

I *Closing Costs* — Amount is the total of the real estate commission fees, recording fees, title fees, and any other credits given to the buyer at the close of escrow for the sale of the property. This is also a negative number because the seller won't receive this money; it's been taken out of the funds to be disbursed to the seller (contractor).

I *Const Draw* — This is the amount the escrow company paid to the lender for the total amount of construction draws taken to date. This total will be listed on the closing papers as the loan payback. It's a negative number because it's being taken out of funds to be disbursed to the seller (contractor). This line pays back the construction loan.

The total amount of the cash sale needs to be exactly equal to the amount of money received at the close of escrow.

Recording a Title Company Payment for the Sale of a Property

To record the payment from the title company for the sale of a property:

▌ Click **Home**, then click **Receive Payments**, or from the **Customers** menu, choose **Receive Payments**.

▌ From the pull-down **Received From** list, select the same customer that's on the invoice for the sale of the property.

▌ In the list, select the invoice you received payment for.

▌ Fill in the date you received the payment, the amount paid, the payment method, and the check number if paid by check.

▌ Click **Save & Close**.

To record the payment after depositing it in the bank:

▌ From the **Banking** menu, choose **Make Deposits**.

In this chapter, you've learned how to set up QuickBooks Pro to track each transaction using proper accounting methods for a spec builder. This chapter wasn't meant to be read just one time. In fact, this book wasn't meant to be read only one time. We hope you'll use this chapter and the whole book as a reference guide. For example, you can use this chapter to help set up and link items to track WIP and construction loans. When you sell a property, you'll need to refer back to this section and follow the instructions on how to record the sale of the property. When you receive the money, you can go through receiving the payment. Then when you deposit the money to the bank, you can follow our instructions again.

Conclusion

It's rare that we run across a contractor or builder who enjoys accounting, let alone job costing, payroll, workers' comp reports, or doing income taxes. If you enjoyed these things, you'd be an accountant, bookkeeper, or consultant (as we are). But if you've followed the instructions in this book, you're on your way to having an efficient and accurate accounting and job costing system. And the bottom line is the bottom line — with accurate reporting, you'll be more profitable.

By now you realize that setting up QuickBooks Pro isn't an easy task. We've tried, however, to break up the tasks into manageable units that are simple to understand and duplicate. If you followed our directions in setting up QuickBooks Pro, and were diligent in your efforts to enter transactions accurately, you'll be rewarded with clear and useful job cost reports — the Holy Grail of construction accounting. Of course, there are other important accounting reports you can get from QuickBooks Pro. And we encourage you to continue to explore them as they apply to your business. Craftsman

Book Company (www.craftsman-book.com) publishes several "accounting for contractors" books, including *Markup & Profit* and *Builder's Guide to Accounting*, which will help you apply wisdom to your numbers. There are also several on-line sources you can use to increase your construction accounting knowledge. For nonspecific accounting problems, Intuit, the creator of QuickBooks Pro, has its own Web site (www.intuit.com). For QuickBooks Pro help specific to the construction business, take a look at www.onlineaccounting.com.

As you work with QuickBooks Pro, you'll be amazed at its flexibility and power. We don't suggest that you turn a blind eye to new solutions. No one has all the answers, including us. We're constantly finding new solutions and "work-arounds" to make the process smoother. Also, everyone puts different emphasis on the value of individual reports, so one solution may not sit right with everyone. Remember, if the information is what you need, use it. Don't be misled by "pundits" that are selling the "right way."

In writing this book, we've tried to create a system or process that you can follow on a daily basis. We urge you to view accounting and job costing as a repeatable procedure that will get simpler as you do it over and over again. Repeatability works for you in several ways. It cuts down on mistakes and helps you refine procedures for maximum productivity. It also makes it easier to train a new employee.

We also want to touch a little on the future of computerized construction accounting. There's no doubt in our minds that in the future, the Internet will play a major role in your business. You'll be estimating, purchasing, and paying bills on-line — it's not a question of if, but when. And Intuit is at the forefront of on-line accounting initiatives. This type of "e-commerce" will only get easier to do, cheaper to implement, and more popular. The paperless transaction is already here. For example, you no longer get a receipt when you charge something on your credit card over the phone. The construction trades are sometimes slow to catch on, but nevertheless, it won't be long before invoices from your subs and suppliers will go directly into QuickBooks Pro without you ever receiving a paper invoice. And they'll even be earmarked with job costing codes so one day you'll turn on your computer and simply ask for a job cost. Without any input — the data will simply be there. Until then, we'll sit at our computer terminals, wishing we didn't have to code this invoice or break down that timecard.

Estimating with QuickBooks Pro

In Chapter 12 we suggested that any contractor who has a good estimating program and likes it should stick with that program rather than switch to QuickBooks Pro estimating. You can create estimates outside of QuickBooks Pro and enter only section summaries (for example, the job phases in *sample.qbw*) into QuickBooks Pro. We know that works because we know many contractors who do exactly that. But you need to know that there's another way. In this appendix, we'll explain how to use the estimating functions built into QuickBooks Pro.

There's a lot to like about QuickBooks Pro estimating. It works. It comes free with QuickBooks Pro. And, like all QuickBooks, it's very forgiving. Practice all you want. Don't worry about making mistakes. It's easy to undo almost anything.

Three Good Reasons to Try QuickBooks Pro Estimating

First, QuickBooks Pro estimating is the starting point for all the accounting and job cost tracking features built into QuickBooks Pro. Once you've created an estimate, click to turn that estimate into an invoice. The invoice amount becomes an entry in your accounts receivable and an asset on your balance sheet. You can compare the actual cost of doing each part of the job with your estimates. The difference between costs and receipts shows up as your profit. That's a good, professional way to do business. Every construction company should be run that way. And it all starts with a QuickBooks Pro estimate.

Second, the progress billing features in QuickBooks Pro can be a huge advantage if you send out detailed bills for the work done to date and have to keep track of what's been paid and what's still collectible. The only way to do progress billing in QuickBooks Pro is to start with a QuickBooks Pro estimate.

Progress billing straddles two topics, estimating and billing. We're not going to say much more about progress billing in this appendix. Actually, there isn't much to say. The estimating part of progress billing is easy to learn. You'll find more information on this in Chapter 13, Receivables.

Third, your invoice, like your bid or estimate, should identify each cost in the job. The best invoice for construction work is a perfect image of the bid, repeating each item in the bid and showing the same costs. Why create an estimate and then retype a separate invoice? That's duplicate work. QuickBooks Pro converts estimates to invoices at the click of your mouse button.

Most of what we're going to explain in this chapter isn't described in the QuickBooks Pro User's Guide. But if estimating is important to you, keep reading. We're going to cover it all.

A Road Map to Your Destination

You probably know already that QuickBooks Pro doesn't come with any construction cost estimating data. And for good reason. QuickBooks is written to work as well for a dry cleaning business, or a barber shop or a car wash, as for a construction contractor. No wonder you won't find the cost of hanging drywall anywhere in QuickBooks Pro. You have to create every labor and material cost estimate in your database by typing it into the Item List.

Building that Item List manually takes time. It's no problem to type a few dozen descriptions and costs into the QuickBooks Pro Item List. But an Item List suitable for detailed labor and material cost estimates might require several thousand lines. Even after building a long Item List, you may not be satisfied. Finding a particular item on a list of thousands can take a lot of scrolling up and down. There's no Item List index in QuickBooks Pro. You can sort items alphabetically. But items are only alphabetical within each item type, such as Non-inventory Part, Service, or Other Charge.

And there's one important limit built into the QuickBooks Pro Item List. An item name can't be longer than 31 characters. This truncating can be a handicap when you're describing something like "5/8-inch type X gypsum wallboard screwed in place, taped, textured and painted two sides." That's 88 characters, much too long for a QuickBooks Pro item name. So you have to abbreviate, such as GWB5/8XSTTP2S. Not many estimators are eager to deal with several thousand codes like that.

In Appendix B we'll show another way, creating an estimate outside of QuickBooks Pro (using the National Estimator program and ready-made database on the CD in the back of this book) and importing the completed estimate into QuickBooks Pro. The advantages should be obvious. But please don't skip Appendix A just because you prefer doing what's described in Appendix B. Every contractor using QuickBooks Pro should understand QuickBooks Pro estimating. Walk before you run. When you're comfortable with the information in this appendix, move on to Appendix B.

Setting Preferences for Estimating

To begin, let's be sure we're reading off the same sheet by checking your preference settings. See Figure A-1.

▌ From the **Edit** menu, choose **Preferences**.

▌ Click the **Jobs & Estimates** icon and the **Company Preferences** tab.

▌ Answer **Yes** to the question **Do You Create Estimates?**

▌ For now, answer **No** to **Do You Do Progress Invoicing?**

▌ Click **OK.**

▌ Look down the Preferences list on the left side of the window. Click the **Sales Tax** icon and then the **Company Preferences** tab.

▌ Be sure to answer **Yes** to **Do You Charge Sales Tax?** See Figure A-2.

If you collect sales tax and haven't set up the most common sales tax and its rate, you'll need to do that now.

▌ From the pull-down Most common sales tax list, select **Add New**.

▌ In the New Item window, enter the type of tax (probably Sales Tax Item), name of the tax, description, tax rate, and taxing agency.

▌ If you haven't set up the taxing agency, you'll need to do that now. From the pull-down Tax Agency (Vendor) list, select **Add New**.

▌ In the New Vendor window, enter information about the taxing agency.

▌ Click **OK** in the New Vendor window to return to the New Item window.

▌ In the New Item window, if you've entered all necessary information, click **OK**.

Back in the Sales Tax Preferences window:

▌ Don't select Mark taxable amounts with "T" when printing unless you want the taxable amounts on estimates marked with a T in the Tax column.

▌ Click **OK.**

Figure A-1

Use the Company
Preferences tab
of the Jobs & Estimates
Preferences window to
tell QuickBooks Pro
you want to create
estimates.

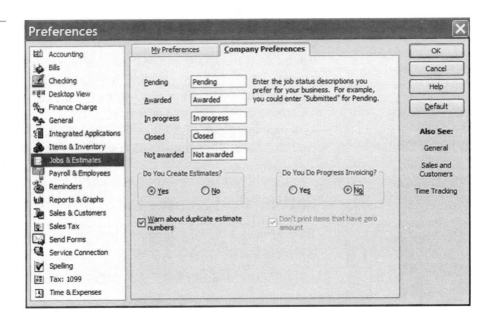

Figure A-2

Enter sales tax
information in
the Company
Preferences tab
of the Sales Tax
Preferences window.

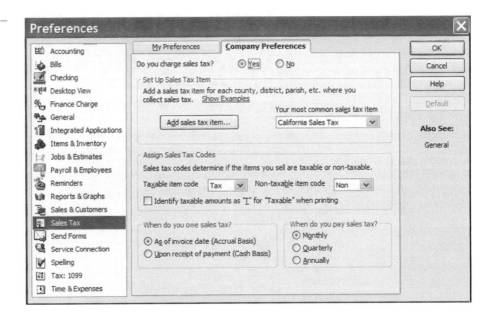

Figure A-3

Use the Create Estimates window to enter a description, quantity, and cost for each item you're including in an estimate.

Building Your Item List

Figure A-3 shows a small estimate for flooring work that's part of a kitchen addition job. Let's take a closer look at this estimate. Then we'll actually create the estimate. In the Item column on the estimate, you'll see three items — Vinyl Tile, Install Vinyl Tile, and Tile Flooring. Since QuickBooks Pro doesn't come with any cost estimating data, we'll begin by adding these items to our Item List.

We recommend that you distinguish between taxable and non-taxable items in your Item List. In many states, the materials you install are taxable to your customer. There's usually no tax on labor. If you collect tax from your customers and have to remit that tax to a state agency, your estimates and invoices have to identify what's taxable and what's not. Some states and Canadian provinces require that you keep a record of the tax collected on each invoice and write a check to the taxing authority when tax payments are due. If your state places a tax on material alone, you have to keep material costs separate from labor costs on your Item List. Keep that in mind as we create new items in the next section.

Even if you don't have to charge sales tax on material, it's good practice to separate labor and material costs on your Item List. That lets you compare estimated material costs and estimated labor costs with actual material costs and actual labor costs. Of course, for subcontracted work, you have only one lump sum — including both material and labor. There's usually no reason to separate labor and material costs for subcontracted work.

Figure A-4

Use the New Item window to enter a non-inventory part type item.

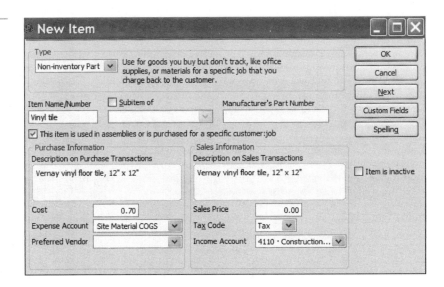

Creating a New Item to Use on Estimates

There are two ways to build an Item List and we'll cover both. You can create items before you use them on an estimate. We'll demonstrate this first. Or you can create items as part of the estimating process. We'll demonstrate that when we begin writing the estimate.

To add a new item to the Item List:

▌ Click **Home**, then click **Items & Services**, or from the **Lists** menu, choose **Item List**.

▌ In the Item List window, click **Item** and select **New**.

▌ From the pull-down Type list, select **Non-inventory Part**. QuickBooks Pro doesn't have a "material" category on the Item List. "Non-inventory Part" is as close as we're going to get. For subcontract costs, or if you prefer combining labor and material costs into one lump sum, select **Other Charge** from the pull-down Type list. We recommend using the Service category for labor costs exclusively.

▌ In **Item Name/Number**, enter the name of the new item. We entered Vinyl Tile as shown in Figure A-4.

▌ Select **This item is used in assemblies or is purchased for a specific customer:job**.

▌ In **Description on Purchase Transactions**, enter an appropriate description. We entered Vernay vinyl floor tile, 12" x 12". What you enter will automatically appear under Description on Sales Transactions, when you press tab.

▌ In **Cost**, enter the cost of the item. Our cost per tile is $.70.

▌ From the pull-down Expense Account list, select any Cost of Goods Sold account that makes sense to you. We used Site Material COGS.

- Leave Sales Price at 0.00.

- Select **Tax Code** if necessary. In our example, the vinyl is taxable.

- From the pull-down Income Account list, select any income account that makes sense to you. We used Construction Income.

- Click **OK**. Then check the spelling if you want.

A few words about what we've just done.

First, we've entered a cost price but not a sales price. You don't have to enter a sales price. That's optional. You may prefer to leave the price open until you use the item in an estimate. Your selling price can change a lot. It's usually easier to enter the correct price when you use the item in an estimate than it is to keep changing prices on the Item List.

Second, you can enter both a cost and a selling price for an item. The difference between cost and selling price is markup, and markup usually varies by job, not type of material. That is another reason we recommend leaving the sales price at 0.00 when you create a new item on the Item List.

Third, we've selected Non-inventory Part as the item type for the Vinyl Tile. If you have an inventory of vinyl tile and want to keep track of how much is in stock, there's another choice. Select Inventory Part as the item type if you intend to record each addition to inventory and sale from inventory. QuickBooks Pro will keep track of your tile inventory. But first you have to set the Purchases & Vendors Preference to "Inventory and purchase orders are active." For more information on preferences, see Chapter 2.

Since every cost item in your estimate has to be on your Item List, you'll get a warning if you try to enter an item that's not on the list. But that's not as inconvenient as you might expect. QuickBooks Pro lets you set up the item without leaving the estimate. We'll demonstrate this shortly.

Create another item for "Install Vinyl Tile." This is labor and should be entered as a "Service" type item. We'll create a subtotal item for "Tile flooring" during the estimating process to demonstrate how you could set up a new item while writing an estimate.

Creating an Estimate in QuickBooks Pro

Let's assume that you've set up the labor and material costs you use most often. Now it's time to start an estimate:

- Click **Home**, then click **Estimates**, or from the **Customers** menu, choose **Create Estimates**.

- From the pull-down **Customer:Job** list in the Create Estimates window, select the customer and job. If you need help in entering a new customer, see Chapter 7, Customers and Jobs.

Figure A-5

Use the New Item window to enter a subtotal type item.

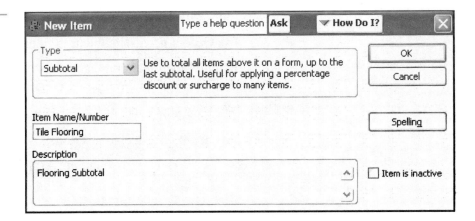

> ▌ Click in the **Description** column and type a short job title, such as "Addition to Kitchen." Don't press Enter unless you intend to insert a blank line below the description you just typed. Then describe the work you plan to do. Refer back to Figure A-3.

> ▌ On the line below the description you just entered, click in the **Item** column, and from the pull-down Item list, scroll down until you find the item you need. In our example, we selected Vinyl Tile. Description and price will fill in automatically.

> ▌ In **Qty**, enter the quantity. Amount will fill in automatically. In our example, we entered 225.

> ▌ In **Markup**, enter the markup. In our example, it's 70%.

> ▌ Use the same procedure to add another item for Install Vinyl Tile. Use the same quantity and markup.

Adding a New Item While Creating an Estimate

Here's how to add a new item to the Item List as you create an estimate:

> ▌ Click in the **Item** column just below the line "Install Vinyl Tile."

> ▌ Click the down arrow and from the pull-down list, select **Add New**.

> ▌ In the New Item window, from the pull-down Type list, select **Subtotal**.

> ▌ Enter **Item Name/Number** and **Description**. In our example shown in Figure A-5, we entered Tile Flooring and Flooring Subtotal.

> ▌ Click **OK** to return to the Create Estimates window.

Figure A-6

Use the New Item window to select items to group together.

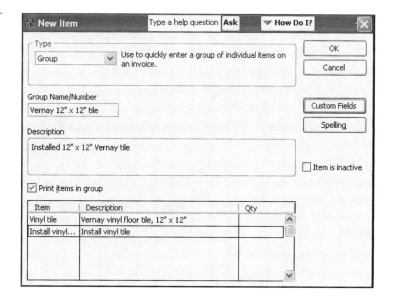

Grouping Labor and Material Costs

Once you have both labor and material items for some tasks on the Item List, consider grouping the two together so you can use them as one item on an estimate. Let's bundle the vinyl tile material and tile installation labor to create a group we'll call Vernay tile.

▌ In the Create Estimates window, from the pull-down Item List in the **Item** column, select **Add New**.

▌ In the New Item window, from the pull-down Type list, select **Group**.

▌ In **Group Name/Number**, enter the group name. In our example shown in Figure A-6, we entered Vernay 12" x 12" tile.

▌ In **Description**, enter the description of the group item. In our example, we entered Installed 12" x 12" Vernay tile.

▌ Click **Print items in group** if you want all the items in the group printed on an estimate.

▌ In the **Item** column, select each item to include in the group from the pull-down Item List. In our example, we selected Vinyl Tile and then Install Vinyl Tile.

▌ Click **OK** to complete the group and return to the Create Estimates window.

To use a group item:

▌ In a Create Estimates window, from the pull-down Item List, select the group item. Note that you still have to enter a quantity for each item in the group.

Item groups are most useful when many components are used together. For example, a group "Wall assembly" might include studs, plates, blocking, nails, 1/2-inch gypsum wallboard, paint, and the associated labor cost. Unfortunately, a group can't include another group, so you have to create your groups one at a time.

Adding a Tax Due Item in a QuickBooks Pro Estimate

Handling tax is another place where QuickBooks Pro really shines. By default, taxable items will have the word Tax in the Tax column at the right side of the estimate. Looking back at Figure A-3, you'll see Tax in the row Vinyl Tile. The tax total of $20.75 for the job is at the lower right of the estimate.

QuickBooks Pro will keep track of sales tax due in each state or county where you do business. If you need to, you can create an item on your Item List for each state or county where you file a sales tax report. To do this:

▌ From the pull-down Tax list near the bottom of the Create Estimates window, select **Add New**.

▌ Enter the tax name, description, rate, and tax agency.

▌ Click **OK** to return to the Create Estimates window.

The tax due will be calculated automatically for each taxing authority you set up. Of course, you have to select the correct taxing authority when creating estimates. When the time comes to pay the tax, choose the Pay Sales Tax command from Sales Tax on the Vendors menu to see how much you owe each taxing authority.

Our sample estimate is now complete:

▌ In the Create Estimates window, click **Save & Close**.

Turning an Estimate into an Invoice

QuickBooks Pro estimates don't have any effect on account balances. You can do all the estimates you want without changing anything on the company profit and loss statement or balance sheet. That's no longer true when you turn an estimate into an invoice. Estimates are just talk. Invoices are real money. To turn an estimate into an invoice:

▌ Click **Home**, then click **Estimates**, or from the **Customers** menu, choose **Create Estimates** and select the estimate. We'll select our flooring estimate in Figure A-3.

▌ Click the **Create Invoice** icon at the upper part of the window.

▌ Click **OK** and you'll see the invoice. Figure A-7 shows our invoice.

Notice that the Markup column is gone. In an invoice, the Amount column includes markup. So the balance due on an invoice will be the same as the estimate total. Only the way we arrived at that total has changed.

Figure A-7

An invoice created from an estimate.

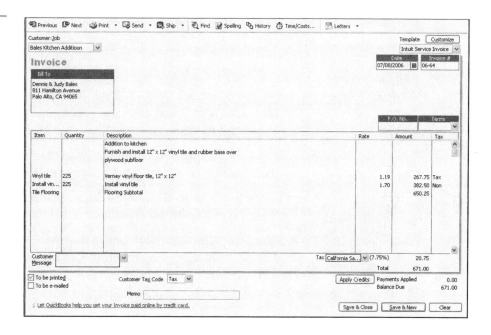

Let's take a tour around the invoice. You can change anything now. The invoice isn't official until you click on Save & Close.

▌ At the upper center of the Create Invoices window, consider changing Form Template to Intuit Service Invoice. That's more appropriate for a construction contractor.

▌ Check to be sure the Invoice #, Date, P.O. No., and Terms are what you want.

▌ In the memo box at the lower left of the estimate, type a short message if you want to.

▌ Check all the prices and descriptions to be sure they're correct. You can change invoice prices, quantities, or descriptions here. You can even add or delete lines at this point. Click the line on the estimate that you want to change or where you want to insert a line. Then click on **Edit** and Insert Line or Delete Line.

▌ When the invoice is perfect, print it. Then click **Save & Close**. The Create Invoices window will close after offering to check the spelling on the invoice.

When you click Save & Close in the Create Invoices window, you make major changes throughout your accounting system. Remember, an invoice represents money, and the primary job of QuickBooks Pro is to keep track of money. Let's look at some of the changes you made by clicking Save & Close in the Create Invoices window.

The Invoice in Your Accounting System

Totals from the new invoice will show up throughout QuickBooks Pro reports. You can check these reports to see for yourself.

▍ From the **Report Center** or the **Reports** menu, choose **Company & Financial**. The Profit and Loss reports show the income. The Balance Sheet reports show Accounts Receivable and the sales tax payable.

▍ From the **Report Center** or the **Reports** menu, choose **Customers & Receivables**. The A/R reports show the receivable.

▍ From the **Report Center** or the **Reports** menu, choose **Sales**. The Sales reports show the sale.

▍ From the **Report Center** or the **Reports** menu, choose **Jobs, Time & Mileage** and then **Jobs Estimates vs. Actuals Detail** to show the cost and revenue.

That one little invoice has worked its way throughout your QuickBooks Pro company data file.

Now, here's the part we like. QuickBooks Pro is so forgiving. Since we were just practicing with the Bales job, let's delete the invoice and see what happens.

▍ Close any reports you have open.

▍ Click **Home**, then click **Create Invoices**, or from the **Customers** menu, choose **Create Invoices** and then **Previous** to get back to the Bales invoice.

▍ With the Bales invoice on your screen, click **Edit** in the menu bar and **Delete Invoice**.

▍ Click **OK** and the invoice is gone, not just off your screen, but throughout your QuickBooks Pro company data file. There's no trace of the Bales invoice anywhere in the QuickBooks Pro reports.

▍ Click **Home**, then click **Estimates**, or from the **Customers** menu, choose **Create Estimates** and click **Previous** to search for the Bales estimate. You'll find it's still there. Deleting the invoice had no effect on the estimate. That means you could recreate the invoice once again from the estimate if necessary.

That's only part of the power in QuickBooks Pro. Any time before the invoice is paid, you can add lines, delete lines, or change anything you want. That's a nice feature. But, as a practical matter, once an invoice has been mailed to a client, it's better to issue a credit or create another invoice if you have a lot of changes to make.

Figure A-8

Use the Clean Up Company Data window to select transactions you want QuickBooks Pro to remove.

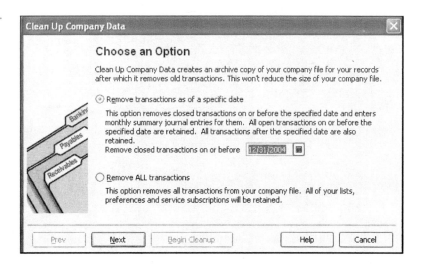

Tidying Up Your Company File

When is it time to delete an invoice or estimate? Don't delete any estimate just because you've turned it into an invoice. If you delete an estimate too soon, you'll remove all estimated costs from the Job Estimates vs. Actuals reports. You need that information at least until you close the job and pay all the bills. You can usually delete any estimates that weren't accepted. That'll help reduce clutter in your company file.

Keep invoices (even after paid in full) at least until the end of the fiscal year during which there was any activity.

Here's the best rule to follow. Don't delete anything in your QuickBooks Pro company data file just because it's old and (you hope) no longer needed. Better to let QuickBooks Pro decide what isn't needed.

Here's how to clean up a company file:

▌ From the **Window** menu, select **Close All**.

▌ From the **File** menu, select **Utilities**, then select **Clean Up Company Data...**

▌ In the Clean Up Company Data dialog box, select **Remove transactions as of a specific date**. See our example in Figure A-8.

▌ Enter the date. It's good practice to save data in the current year and prior year.

▌ Follow instructions on the screen.

That's about all there is to know about estimating with QuickBooks Pro. In Appendix B, we'll go into more depth and show you how to prepare detailed estimates using National Estimator (a program on the CD in the back of this book) and then import those estimates into QuickBooks Pro.

Job Cost Tracking and Importing Estimates

A Different Approach

This appendix is different from what you find in the first 19 chapters. Until now, we've covered the basics. And we've steered clear of topics some could find confusing or frustrating. This appendix is intended more for specialists, construction contractors who:

▌ Consider construction cost estimating an important task.

▌ Need a database with hundreds of pages of construction costs estimates.

▌ Write estimates that include dozens or hundreds of cost items.

▌ Want to compare estimated costs and actual costs when the job is done.

▌ Don't want to re-type their estimates into QuickBooks Pro.

Appendix A covered estimating as it exists in QuickBooks Pro. A major disadvantage is that QBP doesn't include cost estimating data. You have to build the database and keep that data current. There's another choice, and that's the subject of this Appendix B. The disk in the back of this book has several hundred pages of labor and material costs for construction, plus a program that makes it easy to compile estimates using this data. Once you've built an estimate, it exports directly to the estimating module in QuickBooks Pro, creating an Item List automatically. That's cool. It can save hours on any estimate.

Hundreds of construction cost estimating programs are available. A few export estimates to QuickBooks Pro. But none is used more widely than Craftsman's National Estimator, the program on the disk in the back of this manual. About a dozen estimating databases work with the National Estimator program. All are revised annually and cost about $60, including the entire database as a printed book. An order form is at the back of this manual. If you don't need a printed book, all Craftsman estimating databases are available by Web download for much less than $60. Point your browser at http://craftsman-book.com. These downloads are guaranteed to work perfectly on your computer.

Estimates created with any version of National Estimator export smoothly to QuickBooks Pro. That's what we're going to explain in this chapter — step-by-step.

Tracking Job Costs

When you use QuickBooks Pro to write a check for a construction expense (payroll, supplies, subcontracts), you can charge the expense to some cost category in the job estimate. See the five columns in the Estimates vs. Actuals Detail report for the Palmer Job on the next page.

▌ *Column 1 (Est. Cost)* — Estimated costs by job phase for material, labor, equipment, and subcontract expense.

▌ *Column 2 (Act. Cost)* — Actual cost to date for each of those categories.

▌ *Columns 3, 4 ($ Diff., % Diff.)* — The difference between estimated and actual cost (both in dollars and percent).

▌ *Column 5 (Act. Revenue)* — Actual revenue to date (the amount invoiced).

If you need information like this, you can use QuickBooks Pro to simplify collecting and recording job expenses so you get detailed reports from the beginning to the end of every job.

Why track job costs? Easy. There should be two profits in every job. The first is the money that ends up in your pocket. The second is what you learned. If you don't keep detailed cost records, you're missing part of the profit available on every contract. You're not finding the financial sinkholes that can turn any good job into a financial disaster. More likely, you'll keep making the same estimating mistakes over and over — until someone (your lender, your employees, your family) suggests that you get into another line of work.

What Is Job Cost Tracking?

Job cost tracking (sometimes called job costing) is one of QuickBooks Pro's major strengths. In fact, QuickBooks Pro is the only modestly-priced accounting program we know that does real job cost tracking. Every check you write for construction work — to suppliers, subcontractors, and employees — gets charged to a cost category in an estimate. Nothing sneaks by. You can see at a glance if costs are in line with the original bid.

Robins Construction
Job Estimates vs. Actuals Detail for Alan Palmer
All Transactions

	Est. Cost	Act. Cost	($) Diff.	(%) Diff.	Act. Revenue
Parts					
DemolitionMat ▶	2,001.00 ◀	1,843.67	-157.33	-7.9%	2,001.00
ElectricalMat	20.00	21.00	1.00	5.0%	20.00
FlooringMat	1,240.00	1,196.47	-43.53	-3.5%	1,240.00
FoundationMat	1,890.13	2,046.56	156.43	8.3%	1,890.13
Int finishMat	1,668.50	3,217.95	1,549.45	92.9%	1,668.50
PaintingMat	966.24	897.42	-68.82	-7.1%	966.24
PlumbingMat	1,231.40	1,341.43	110.03	8.9%	1,231.40
Rough framMat	7,022.20	6,841.12	-181.08	-2.6%	7,022.20
Total Parts	16,039.47	17,405.62	1,366.15	8.5%	16,039.47
Service					
DemolitionLab	9,326.40	9,470.70	144.30	1.5%	9,326.40
FlooringLab	200.88	189.00	-11.88	-5.9%	200.88
FoundationLab	4,052.35	3,914.60	-137.75	-3.4%	4,052.35
Int finishLab	3,186.77	5,879.73	2,692.96	84.5%	3,186.77
PaintingLab	927.09	874.50	-52.59	-5.7%	927.09
PlumbingLab	724.90	943.00	218.10	30.1%	724.90
ProjectLab	1,250.00	1,250.00	0.00	0.0%	1,250.00
Rough framLab	5,301.94	4,940.00	-361.94	-6.8%	5,301.94
Total Service	24,970.33	27,461.53	2,491.20	10%	24,970.33
Other Charges					
ElectricSub	1,797.50	1,890.00	92.50	5.1%	1,797.50
FlooringSub	1,845.00	1,903.11	58.11	3.1%	1,845.00
FoundationEqu	1,282.56	1,100.00	-182.56	-14.2%	1,282.56
Overhead	4,593.49	4,500.00	-93.49	-2%	4,593.49
Profit	7,579.25	0.00	-7,579.25	-100.0%	7,579.25
Total Other Charges	17,097.80	9,393.11	-7,704.69	-45.1%	17,097.80
TOTAL	**58,107.60**	**54,260.26**	**-3,847.34**	**-6.6%**	**58,107.60**

Tracking costs on the Palmer job.

All the big engineering and construction companies have, and use, job cost tracking programs — usually costing thousands of dollars. Now you have one, too. If you haven't seen a job costing program in operation, keep reading. You may find this very interesting.

If your construction company estimates labor, material and equipment costs, and you have several crews working several jobs — or even one large job — you need to track job costs.

Here's an example. A contractor we know got a contract to build a large custom home. Most of the work was to be done with his crews. But 11 subcontractors would be required. Once work got started, invoices, purchase orders, checks, and bills were received and sent out nearly every day. This contractor's job cost system consisted of some numbers he scratched out with a stick on the dusty bed of his pickup truck. When material invoices for the foundation and framing came in, he paid the bills, not realizing that the cost was nearly 20 percent higher than quoted. A job cost tracking system would have identified the problem immediately. He should have called the vendors to complain. But he didn't do the math before paying the bills and didn't realize he'd been overcharged. Weeks later, when trying to figure out why he lost money on the project, he pulled out some old paperwork for the job and discovered the overcharges. By then it was too late.

If this example sounds familiar to you, it's time to start tracking costs on your jobs.

Estimates into Invoices

In Chapter 12 and Appendix A we mentioned that QuickBooks Pro estimating might not be the best choice for all contractors. The National Estimator program on the disk in the back of this book includes Job Cost Wizard, a bridge that exports completed estimates to QuickBooks Pro so you can use the accounting and tracking capabilities in QuickBooks Pro.

If you dream of turning estimates into invoices at the click of a mouse button, we'll show you how. If you need help, call the Craftsman Book Company support line at 760-438-7828 from 8 AM to 5 PM (Pacific) Monday through Friday. Help is free. All you pay for is the phone call.

We should let you know about one drawback to using our method to turn an estimate into an invoice. When you do this, you'll see that the material, labor, equipment, and subtotal costs in your estimate become billable items on the invoice. Maybe you don't want clients to see that much detail. Maybe you'd prefer to show just a lump sum cost for each job phase. We understand. Lump sum invoicing should be an option in QuickBooks Pro. Unfortunately, it's not. If you want to track labor, material, and subcontract expenses by job, you have to show labor, material, and subcontract expenses in the invoice. If you agree that tracking costs by phase (foundation, framing, finishing, etc.) and by type (labor, material, and subcontract) is valuable, you're probably ready to tolerate invoices that show too much detail. And there's another reason to identify material costs on your invoices — a little matter of sales tax.

Handling Tax

If you look back through the first 19 chapters of this book, you'll discover a significant omission. Tax doesn't appear anywhere. Why is that? Isn't tax important? Doesn't QuickBooks Pro handle tax? Are we suggesting that contractors should ignore sales tax?

Hardly! Unless all your work is for tax exempt government agencies, tax is an important cost in your jobs, usually 6 to 8 percent of material expense. But notice the word *usually*. Sales tax varies by state and province. And even within states, sales tax rates vary — not only by amount, but also by what gets taxed, when it gets taxed, and who has to collect and remit taxes to the taxing authority. Rules vary among the 50 states. And many states have dozens of taxing authorities, each with the right to set rates and decide what gets taxed and what doesn't. There's no uniform national rule so there's nothing we can say that applies everywhere.

However, QuickBooks Pro was developed for retail businesses to use, so it does an excellent job identifying what's taxable and what isn't, recording taxable sales, and even remitting the tax collected to the government authority.

True, some contractors can almost ignore sales tax. "Paper contractors" who subcontract everything in the job can ignore sales tax in most states. Connecticut is an exception. All of Canada is another exception. In Canada, contractors with sales volume over $30,000 have to add federal GST (Goods and Services Tax) to invoices and remit at least quarterly to the federal government. The only contractors in Canada that can ignore sales tax are contractors who work on government jobs exclusively.

Only a few states and provinces impose sales tax on labor. Only a few states and provinces require construction contractors to charge sales tax and remit the amount collected to a government agency. But nearly all construction contractors have to pay tax on at least some materials they purchase for installation. That makes tax an important cost component. So if you're going to pay tax on materials, you'll want to include tax in your material cost estimates. And therein lies the problem when we begin converting estimates to invoices.

How many of your material suppliers show shelf prices that include sales tax? For example, call for a quote on framing lumber. Does that quote include sales tax? The price posted on gasoline pumps includes tax. But we've never seen construction materials priced with tax. The tax is understood, not expressed. So, unless you fall under one of the exemptions listed above, you're going to pay sales tax on materials. Most general contractors learn very quickly to add sales tax on materials in their estimates. They have to.

If you use QuickBooks Pro to convert an estimate into an invoice, the invoice has to mirror sales tax in the estimate. If you pay tax on materials purchased for a job, you can mark each material subtotal on the invoice as taxable and accumulate a total for tax at the bottom of the invoice. Neat and easy! You pay tax on materials and that tax gets passed neatly and efficiently to your customers through invoices. But it only works if you identify material costs in your estimates.

Using Items for Job Cost Tracking

Look back at the three major cost types at the left margin in the Job Estimates vs. Actuals Detail report for the *Palmer* job.

▌ *Parts* includes all material costs

▌ *Service* includes all labor costs

▌ *Other Charges* includes equipment, subcontracts, overhead, and profit

Under each of the three cost types in the report, you see cost categories. For example, the first cost category under Parts is "DemolitionMat." This cost category covers demolition material expense such as tarps, tape, and hauling and dump fees. Demolition labor expense "DemolitionLab" is listed under the second cost type, "Service."

The first part of each cost category name comes from the estimate. The last three letters are either:

▌ *Mat* for material

▌ *Lab* for labor

▌ *Equ* for equipment

▌ *Sub* for subcontract

Cost categories in this estimate are Demolition, Electrical, Flooring, Foundation, Int finish, and so on. The estimator selected these descriptions when he created the estimate. It's OK to use any descriptions you want. You can use some list of "standard" construction phases or names you make up on the spot. Everything works. The 31 construction categories in *sample.qbw* or *company.qbw* are perfect if that's what you want. But they're not required.

Cost category names (the Item List) have to be fairly short. These short category names are fine when printing reports. Short names leave plenty of room on the page for cost columns. But it means you'll have to abbreviate category names at least occasionally when creating estimates.

Check the bottom line of the Palmer job report. The estimated cost is the same as the contract price, $58,107.60. The actual cost was $54,260.26. So the Palmer job was a winner by $3,847.34. Right? Look closer. Interior finish ran $4,242.41 over the estimate ($1,549.45 plus $2,692.96). Where does that money come from? You know already. Straight out of job profit. How long can you afford to wait to find problems like this?

One more question. What if you could get a job cost report like this (and lots more) on all your jobs without doing any extra paperwork, and using software you already have? What if there was a program that created the QuickBooks Pro Items for you automatically. Would you try it? Of course you would. And that's exactly what this appendix is going to explain.

National Estimator icon.

Job Cost Wizard icon.

The Job Estimates vs. Actuals Detail (the Palmer job) is a QuickBooks Pro report. But the estimate we show in the figure was created in National Estimator, converted to QuickBooks Pro format by Job Cost Wizard, and then imported into QuickBooks Pro. You've got National Estimator and Job Cost Wizard on the CD in the back of this book. You also have over a hundred pages of manhour estimates and labor costs for residential and commercial construction.

If you estimate larger jobs and prefer to start with an existing labor and material cost estimating database rather than building your own Item List in QuickBooks Pro, this is the information you need.

Let's import an estimate we made using National Estimator and converted using Job Cost Wizard:

▌ In QuickBooks Pro, from the **File** menu, choose **Utilities** and **Import**. Then click **IIF files...** Verify that you are looking in the QuickBooks Pro folder.

▌ Double-click **Home_off.iif**. When the estimate for this home office job has been imported, you may have to click **OK**. Ignore any warning message about job or estimate information.

▌ Click on **Customers**.

▌ Click on **Create Estimates**.

▌ Click **Previous** until you see *ESTIMATE NO.* 101 for the Stillwel Job.

▌ Use the scroll bar at the right of the estimate to page through the estimate.

Importing an estimate (such as the Stillwel home office job) will set up all the items you need on the Items List. If you click the Create Invoice button in the Create Estimates window now, all costs in the Stillwel estimate will become charges in the Stillwel invoice. You probably won't want to do that.

To exit the Create Estimate window without saving the estimate:

▌ From the **Edit** menu, choose **Delete Estimate**.

▌ Click **OK**.

In this example, the Stillwel estimate had already been created in National Estimator and converted with Job Cost Wizard. But you'll want to create your own estimates. So the first step will be learning National Estimator.

Help Learning National Estimator

Craftsman
ShowMe

ShowMe is an interactive video guide to National Estimator. Sit back and relax and let the interactive video run, or jump from topic to topic. Exit any time you want. Then go back later to brush up on anything you missed. ShowMe takes up very little space on your hard drive. But the Contractor's Guide to QuickBooks Pro disk has to be in your CD drive when watching ShowMe.

There are at least three ways to start ShowMe:

▌ You'll see links to ShowMe on many Help topics when using the Help file.

▌ Click on **ShowMe** in the Construction Estimating program group to run all of ShowMe.

▌ Click on the "**Show me!**" button below Dan's Tip of the Day to get more information about that tip.

Forty Estimating and Bidding Forms are included on the Contractor's Guide to QuickBooks Pro CD. These forms are on the FORMS directory on the CD, and will install on your hard drive in My Documents\Construction Forms. If you have a laser or ink jet printer, you'll be able to create top-quality customized forms in minutes. To use these forms, you'll need any of the popular word processing or spreadsheet programs. First, start your word processing or spreadsheet program. Then click on **File** and **Open**. Browse to the My Documents\Construction Forms folder. You'll see over forty forms created especially for each of the following programs:

▌ Microsoft Word (file name extension of DOC)

▌ Microsoft Excel (file name extension of XLS)

▌ Lotus 1-2-3 (file name extension of WK4)

▌ Microsoft Works (file name extension of WKS)

▌ WordPerfect (file name extension of WP)

To open any form, double-click on the form name. Make the changes you want. Then save the modified form to your hard drive (such as C:). For a description of all forms on the CD, open the file INDEX.TXT on the FORMS directory.

National Estimator Help has everything you could ever want to know about National Estimator. These help files are available any time National Estimator

National
Estimator Help

is running (click on the question mark) or from the Construction Estimating program group (click on **National Estimator Help**). To print a user's guide to National Estimator, click on the question mark. Then click on **Print All Topics**. Click on **File** on the Help menu bar. Click on **Print Topic**. Click on **OK**.

Job Cost Wizard Help explains how to use Job Cost Wizard to convert National Estimator estimates into invoices and then export the estimate or invoice to QuickBooks Pro (version 5 and higher). This help file is available any time you're running Job Cost Wizard. Just click **Help** on the menu bar. Then click on **Contents** and the topic of your choice. From Help Contents you can print a guide to Job Cost Wizard: Click on **Print All Topics**. Then click on **File**, **Print Topic** and **OK**.

System Requirements

National Estimator and Job Cost Wizard require Windows 98 Second Edition or better, including Windows Vista, and about 160 Mb free on a hard drive.

Installing Contractor's Guide

With Windows running, put the CD in your CD drive (such as D:). If installation doesn't start after a few seconds:

1. Click on **Start**
2. Click on **Control Panel**.
3. Double-click on **Add or Remove Programs**.
4. Click on **Add New Programs**.
5. Click on **CD or Floppy**.
6. Click on **Next**.
7. Select **D:\Setup**
8. Click on **Finish**

Follow the instructions on the screen. We recommend accepting the installation defaults. When installation is complete, click on **Finish**. National Estimator will begin automatically.

Uninstalling Contractor's Guide

Click on **Start**, **Settings**, **Control Panel**. Double-click on **Add or Remove Programs**. Then click on the name of the program to remove and **Add or Remove**.

Estimating with National Estimator

National Estimator begins when you click on the National Estimator icon or click on **Start**, **Programs**, the Construction Estimating group and then National Estimator.

On the title bar at the top of the screen you see the program name, National Estimator, and [Construction Estimating Reference Data]. Let's take a closer look at the other information at the top of your screen.

The Menu Bar

Below the title bar you see the menu bar. Every option in National Estimator is available on the menu bar. Click with your left mouse button on any item on the menu bar to open a list of available commands.

Title Bar
Menu Bar
Toolbar
Page & Subject
Column Heads

Buttons on the Toolbar

Below the menu bar you see 24 buttons that make up the toolbar. The options you use most in National Estimator are only a mouse click away on the toolbar.

Column Headings

Below the toolbar you'll see column headings for your costbook:

Craft@Hrs for craft (the crew doing the work) and manhours (to complete the task)

Unit for unit of measure, such as per linear foot or per square foot

Material for material cost

Labor for labor cost

Equip for equipment cost

Total for the total of all cost columns

The Status Bar

The bottom line on your screen is the status bar. Here you'll find helpful information about the choices available. Notice "Page 20/360" near the center of the status line. That's a clue that you're looking at page 20 of a 360-page book. The costbook name is Construction Estimating Reference Data or "CERD" for short.

Check the status bar occasionally for helpful tips and explanations of what you see on screen.

The Costbook

All manhour estimating tables in Construction Estimating Reference Data are available in the Costbook window. Notice also the words Page 20 Sitework at the left side of the screen just below the toolbar. That's your clue that the Sitework section of page 20 is on the screen.

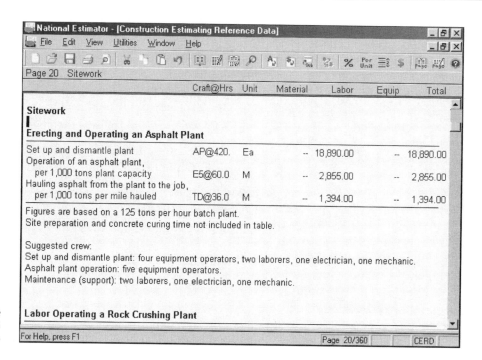

The Costbook Window has all the manhour estimates in Construction Estimating Reference Data.

To turn to the next page, either:

▌ Press [PgDn] (with Num Lock off), -or-

▌ Click on the lower half of the scroll bar at the right edge of the screen.

To move down one line at a time, either:

▌ Press the [↓] arrow key (with Num Lock off), -or-

▌ Click on the down arrow on the scroll bar at the lower right corner of the screen.

Type the page number you want to see.

To turn quickly to any page, either:

▌ Click on the 🖳 (Turn to Costbook Page) button near the right end of the toolbar, -or-

▌ Click on **View** on the menu bar. Then click on **Turn to Costbook Page**.

Type the number of the page you want to see and press [Enter ⏎]. National Estimator will turn to the top of the page you requested.

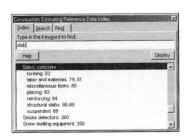

Use the electronic index to find cost estimates for any item.

The index jumps to Slabs, concrete.

A Still Better Way: Keyword Search

To find any cost estimate in seconds, search by keyword in the index. To go to the index, either:

▌ Click on the 🔍 (Index) button near the center of the toolbar, -or-

▌ Click on **View** on the menu bar. Then click **Index Window**.

Notice that the cursor is blinking in the *Type in the Keyword to find:* box on your screen. Obviously, the index is ready to begin a search.

If you prefer a Web-like Search, all costbooks include a Search tab. Type the words you're looking for. Click either:

▌ **Exact Phrase**, or

▌ **All Words,** or

▌ **Any Words**

Then click **Search**.

▌ If you prefer to find every occurrence of a word, click the **Find** tab. Type the word and click **Find Next**.

Your First Estimate

Suppose we're estimating the cost of a concrete slab on grade. Let's put the index to work with a search for slabs. In the *Type in the Keyword to find:* box, type *slab*. The index jumps to the heading *Slabs, concrete*.

▌ Click once on the line *labor and materials: 79, 81* and press ⸢ Display ⸥, -or-

▌ Double-click on that line, -or-

▌ Press ⸢Tab⇆⸥ and the ⸢↓⸥ arrow key to move the highlight to *labor and materials: 79, 81*. Then press ⸢ Display ⸥.

If costs appear on several pages, click on the page you prefer.

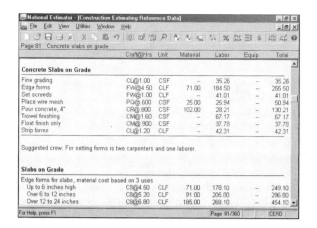

Costs for slabs on grade on page 81.

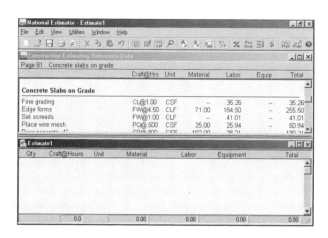

The Split Window: Costbook above and estimate below.

To select the page you want to see (page 81 in this case), either:

▌ Double-click on that line, -or-

▌ Click once on that line and press ⟨Enter ↵⟩.

National Estimator turns to the top of page 81 as shown on the next page.

Splitting the Screen

Most of the time you'll want to see what's in both the costbook and your estimate. To split the screen into two halves, either:

▌ Click on the (Split Window) button near the center of the toolbar, -or-

▌ Click on **View** on the menu bar. Then click on **Split Window** and your screen should look like the example above.

The lines of the costbook on the top half of the screen are for slab base. The bottom half of the screen is reserved for your estimate. Column headings are at the top of the costbook and across the middle of the screen (for your estimate).

To Switch from Window to Window

▌ Click in the window of your choice, -or-

▌ Hold the ⟨Ctrl⟩ key down and press ⟨Tab⇆⟩.

Notice that a window title bar turns dark when that window is selected. The selected window is where keystrokes appear as you type. Click on the bottom half of the screen so your estimate is selected.

Beginning an Estimate

You can type anything in the Estimate Window. Let's start by putting a heading on this estimate:

1. Press Enter↵ once to space down one line.

2. Press Tab⇆ four times (or hold the space bar down) to move the Blinking Cursor (the insert point) near the middle of the line.

3. Type "Estimate One" and press Enter↵. That's the title of this estimate, "Estimate One."

4. Press Enter↵ again to move the cursor down a line. That opens up a little space below the title.

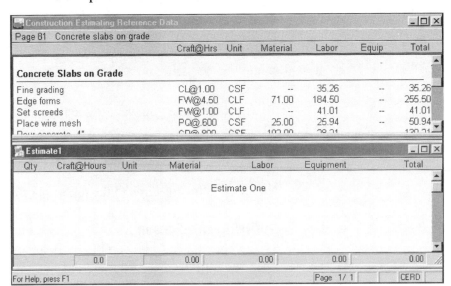

Begin by putting a title on your estimate, such as "Estimate One."

Copying Costs to Your Estimate

Next, we'll estimate the cost of 1600 square feet of fine grading. Click the (Split Window) button on the toolbar to be sure you're in the split window. Click anywhere in the costbook (the top half of your screen). Then press the ↑ or ↓ arrow key (with Num Lock off) until the cursor is on the line:

| Fine grading | CL@1.00 | CSF | -- | 35.26 | -- | 35.26 |

To copy this line to your estimate:

Hint: Instead of clicking on the ▧ (Copy) and ▧ (Paste) buttons, you can both copy and paste by pressing the F8 key.

1. Click on the line.

2. Click the ▧ (Copy) button.

3. Click on the ▧ (Paste) button to open the Enter Cost Information dialog box.

Notice that the blinking cursor is in the Quantity box:

1. Type a quantity of 16 because this job requires 1600 square feet. (The unit of measure is CSF or 100 square feet.)

2. Press ⟨Tab⟩ and check the estimate for accuracy.

3. Notice that the column headed Unit Costs shows labor costs per unit, per CSF (100 square foot) in this case.

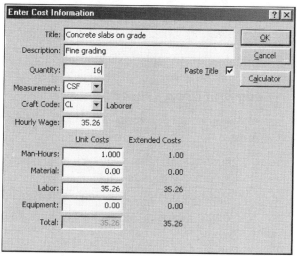

Use the Enter Cost Information dialog box to enter quantities and change costs.

4. The column headed Extended Costs shows costs for the entire 16 CSF (1600 square feet).

5. Press ⟨Enter⟩ or click on **OK** to copy these figures to the end of your estimate.

The new line at the bottom of your estimate shows:

Extended costs for fine grading as they appear on your estimate form.

16.00 is the quantity of fine grading

CL is the recommended crew, one laborer

@16.00 shows the manhours required for the work

CSF is the unit of measure, 100 square feet in this case

0.00 shows that no material cost has been entered in this case

564.16 is labor cost for the job

0.00 shows there is no equipment cost

564.16 is the total of material, labor and equipment columns

Copy Anything to Anywhere in Your Estimate

Anything in the costbook can be copied to your estimate. Just click on the line (or select the words) you want to copy and press the F8 key. It's copied to the last line of your estimating form. If your selection includes costs, you'll have a chance to enter the quantity. To copy to the *middle* of your estimate:

1. Select what you want to copy.
2. Click on the ▧ (Copy) button.
3. Click in the estimate where you want to paste.
4. Click on the ▧ (Paste) button.

Drag and Drop Estimating

You can also drag lines or words out of the costbook and drop them in your estimate:

1. Click on the line to copy and move your mouse slightly so the line is selected (turns black).
2. Release the mouse button. The line remains selected.
3. Click again on the selected line. This time hold your mouse button down.

⊘ *Circle with a bar* 　　　 ▧ *Vertical bar with mouse cursor*

4. Your mouse cursor turns into a circle with a diagonal bar.
5. Holding the mouse button down, move the circle with the bar into the Estimate Window.
6. Once in the Estimate Window, you'll see a vertical bar to the left of your mouse cursor.
7. Move this vertical bar to where the copied line should be pasted.
8. Then release the mouse button. The line is pasted in that position.
9. If the line pasted includes costs, you'll have a chance to enter a quantity.

Right-Click Editing

Right-click editing menu.

Most of what you do in National Estimator is editing — such as Cut, Copy, Paste and Undo. Most editing features have their own button on the toolbar. Of course, all editing functions are available from the Edit selection on the menu bar. But you might find it easier and quicker to open the floating edit menu by right-clicking with your mouse. Press the right mouse button, while in the estimate, to see the menu shown here. All edit functions currently valid will be available. Simply click on the selection you need.

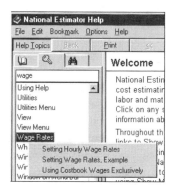

Search information on setting wage rates.

Change the material cost to 1.00.

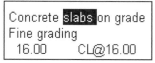

To select, click and hold the mouse button while dragging the mouse.

Changing Wage Rates

The labor cost for fine grading is based on a building laborer working at a cost of $35.26 per hour. Suppose $35.26 per hour isn't right for your estimate. What then? No problem! It's easy to use your own wage rate for any crew or even make up your own crew codes. To get more information on setting wage rates, press [F1]. At National Estimator Help, click on the **Key** (Index) tab. Type *wage* then click on **Wage Rates** and **Setting Hourly Wage Rates**. To return to your estimate, click on **File** on the National Estimator Help menu bar. Then click on **Exit**. Note that the change affects only that single estimate.

Changing Cost Estimates

With Num Lock off, use the [↑] or [↓] arrow key to move the cursor to the line you want to change (or click on that line). In this case, move to the line that begins with a quantity of 16. To open the Enter Cost Information dialog box, either:

▌ Press [Enter ↵], -or-

▌ Click on the [🖫] (Change Cost) button on the toolbar.

To make a change, either:

▌ Click on what you want to change, -or-

▌ Press [Tab ↹] until the cursor advances to what you want to change.

Then type the correct figure. In this case, change the material cost to $1.00.

Press [Tab ↹] and check the Extended Costs column. If it looks OK, press [Enter ↵] and the change is made on your estimating form.

Changing Text (Descriptions)

Click on the [🖾] (Estimate Window) button on the toolbar to be sure you're in the estimate. With Num Lock off, use the [↑] or [↓] arrow key or click the mouse button to put the cursor where you want to make a change. In this case, we're going to make a change on the line that begins "Concrete slabs."

To make a change, click where the change is needed. Then either:

▌ Press the [Del] or [←Bksp] key to erase what needs deleting, -or-

▌ Select what needs deleting and click on the [✂] (Cut) button on the toolbar.

▌ Type what needs to be added.

In this case, click just before "slabs." Then hold the left mouse button down and drag the mouse to the right until you've put a dark background behind "slabs." The dark background shows that this word is selected and ready for editing.

Press the [Del] key, or click on the ✂ (Cut) button on the toolbar, and the selection is cut from the estimate. If that's not what you wanted, click on the ↺ (Undo) button and "slabs" is back again.

Adding Text (Descriptions)

Some of your estimates will require descriptions (text) and costs that can't be found in Construction Estimating Reference Data. What then? With National Estimator it's easy to add descriptions and costs of your choice anywhere in the estimate. For practice, let's add an estimate for four reinforced corners to Estimate One.

Click on the ⊞ (Estimate Window) button to be sure the estimate window is maximized. We can add lines anywhere on the estimate. But in this case, let's make the addition at the end. Press the [↓] arrow key to move the cursor down until it's just above the horizontal line that separates estimate detail lines from the estimate totals. To open a blank line, either:

▌ Press [Enter ↵], -or-

▌ Click on the ᴬ (Insert Text) button on the toolbar, -or-

▌ Click on **Edit** on the menu bar. Then click on **Insert a Text Line**.

Type *Reinforced corners* and press [Enter ↵].

Adding a Cost Estimate Line

Now let's add a cost for "Reinforced corners" to your estimate. Begin by opening the Enter Cost Information dialog box. Either:

▌ Click on the $ (Insert Cost) button on the toolbar, -or-

▌ Click on **Edit** on the menu bar. Then click on **Insert a Cost Line**.

1. The cursor is in the Quantity box. Type the number of units (4 in this case) and press [Tab↹].

2. The cursor moves to the next box, Measurement.

3. From the pull-down Measurement list, select **Ea** (each) and press [Tab↹].

4. Press [Tab↹] twice to leave the Craft Code blank and Hourly Wage at zero.

5. Since these corners will be installed by a subcontractor, there's no material, labor or equipment cost. So press [Tab↹] four times to skip over the Man-Hours, Material, Labor and Equipment boxes.

6. In the Total box, type 20.00. That's the cost per corner quoted by your supplier.

7. Press [Tab↹] once more to advance to OK.

8. Press [Enter ↵] and the cost of four reinforced corners is written to your estimate.

Concrete slabs on grade
Fine grading
　16.00　　　CL@16.00
Reinforced corners

Adding "Reinforced corners."

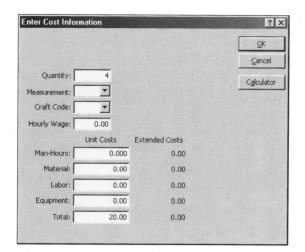

Unit and extended costs for four reinforced corners.

Note: The sum of material, labor and equipment costs appears automatically in the Total box. If there's no cost entered in the Material, Labor or Equipment boxes (such as for a subcontracted item), you can enter any figure in the Total box.

Adding Lines to the Costbook

Add lines or make changes in the costbook the same way you add lines or make changes in an estimate. The additions and changes you make become part of the user costbook. For more information on user costbooks, press F1. Click on the **Key** (Index) tab. Type *add* and click on **Add lines to a costbook**.

Adding Tax

To include sales tax in your estimate:

1. Click on **Edit**.

2. Click on **Current Tax Rates**.

3. Type the tax rate in the appropriate box.

4. Press Tab↹ to advance to the next box.

5. Press Enter↵ or click on **OK** when done.

In this case, the tax rate is 7.25% on materials only. Tax will appear at the end of the estimate.

Type the tax rate that applies.

Subtotals Become QuickBooks Cost Categories

At the end of each section in your estimate, you'll probably want to insert a subtotal. For a general contractor, estimate sections might be Demolition, Excavation, Foundation, Framing, etc. Estimate sections for an insurance repair contractor might include Kitchen, Bathroom or Living Room. Section subtotals help organize your estimates and make them easier to read and understand. Insert section subtotals wherever they make the most sense to you. These subtotals become cost categories when printing bids and invoices. The first 28 characters become cost category names when exporting to QuickBooks.

To insert a subtotal:

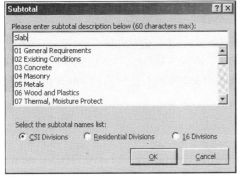

Insert a subtotal.

1. Click on the last cost line of the section (or on any blank line below the section).

2. Click on the ![Subtotal] (Subtotal) button on the toolbar (or click on **Edit** and **Insert or Edit Subtotal Line**).

3. Type a name or description for the section (such as *Slab*).

4. Press ⌈Enter ⏎⌉ or click **OK**.

    If you prefer, select as a subtotal name one of the 16 CSI divisions or one of the 34 residential phases.

Adding Overhead and Profit

Set markup percentages in the Add for Overhead & Profit dialog box. To open the box, either:

▌ Click on the ![$] (Markup) button on the toolbar, -or-

▌ Click on **Edit** on the menu bar. Then click on **Current Markup**.

Type the percentages you want to add for overhead. For this estimate:

Adding overhead & profit.

1. Type *15* on the Overhead line.

2. Press ⌈Tab ⇆⌉ twice to advance to Contingency.

3. Type *5* on the Contingency line.

4. Press ⌈Tab ⇆⌉ twice to advance to Profit.

5. Type *10* on the Profit line.

6. Press ⌈Enter ⏎⌉ or click **OK**.

Markup percentages can be changed at any time. Just click on the ![$] (Markup) button and type the correct figure.

Preview Your Estimate

You can display an estimate on screen just the way it will look when printed on paper. To preview your estimate, either:

▌ Click on the ⬚ (Print Preview) button on the toolbar, -or-

▌ Click on **File** on the menu bar. Then click on **Print Preview**.

A preview of Estimate One.

In print preview:

▌ Click on **Next Page** or **Prev Page** to turn pages.

▌ Click on **Two Page** to see two estimate pages side by side.

▌ Click on **Zoom In** to get a closer look.

▌ Click on **Close** when you've seen enough.

Use buttons in Print Preview to see your estimate as it will look when printed.

Printing Your Estimate

Options available depend on the printer you're using.

When you're ready to print the estimate, either:

▌ Click on the ⬚ (Print) button on the toolbar, -or-

▌ Click on **File** on the menu bar. Then click on **Print Current Estimate...**, -or-

▌ Hold the [Ctrl] key down and type the letter *P*. Press [Enter⏎] or click on **OK** to begin printing.

Save Your Estimate to Disk

To store your estimate on the hard disk where it can be re-opened and changed at any time, either:

▍ Click on the [▣] (Save) button on the toolbar, -or-

▍ Click on **File** on the menu bar. Then click on **Save**, -or-

▍ Hold the [Ctrl] key down and type the letter *S*.

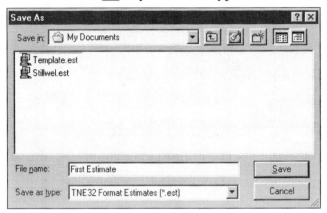

Type the estimate name in the File name box to assign a file name.

The cursor is in the File name box. Type the name you want to give this estimate, such as *First Estimate*. Press [Enter ←] or click on **Save** and the estimate is written to disk. Note that the default location for estimates is the *My Documents* folder.

Select Your Default Costbook

Your default costbook is the last costbook installed. It opens automatically every time you begin using National Estimator. If you have any other costbooks installed on your computer, you can choose which one you want as your default costbook. Usually, it's the costbook you use the most.

Selecting the default costbook.

To change your default costbook, click on **Utilities** on the menu bar. Then click on **Options**. Click the **Select** button within the section titled Default Costbook. Click on the costbook of your choice. Click on **Open**. Then click on **OK**.

Use National Estimator Help

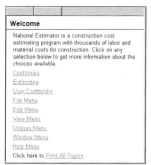

Print a 37-page Guide to
National Estimator.

That completes the basics of National Estimator. You've learned enough to complete most estimates. When you need more information about the fine points, use National Estimator Help. Click on the 🔘 (Help) button to see Help Contents. Then click on the menu selection of your choice. To print instructions for National Estimator, click on **Print All Topics** (at the bottom of the Welcome screen). Click on **File** on the Help menu bar. Then click on **Print Topic**. Click on **Print**.

Converting Estimates with Job Cost Wizard

Use Job Cost Wizard to:

▌ Convert estimates into bids and invoices you can send to a client, and,

▌ Export to QuickBooks where you can track job costs, receivables, payables, create payrolls and print financial reports.

Send the estimate to
Job Cost Wizard.

To view your completed estimate in Job Cost Wizard, either:

▌ Click on **File**, click on **Send Estimate To JCW**, or

▌ Hold the Ctrl key down and tap *J*.

You can also start Job Cost Wizard by clicking on the Job Cost Wizard icon in the Construction Estimating program group. Then click on the name of the estimate you want to open.

The Company Information dialog box will open the first time you use Job Cost Wizard. Type your company name and address. This will appear at the top of every estimate and invoice. When you've filled in information about your company, click on **OK** and your estimate will open in Job Cost Wizard.

Fill in information about
your company.

To change any of the information about your company:

▌ Click on **Options**.

▌ Click on **Your Company Info**.

▌ Type the changes needed and click **OK**.

Zoom, Scroll and Turn Pages

If the estimate doesn't fit your screen, set the percentage of zoom. For 640x480 resolution, type 81% in the zoom window and press [Enter ↵].

Click and drag the vertical slide bar at the right of your screen to scroll down the page. Turn pages by clicking on the Previous Page or Next Page buttons.

Enter Job Information

Job Cost Wizard needs some information about the job to create a nice-looking bid or invoice. For practice, enter job information for the Stillwel estimate:

1. **Customer name and address.** Click on the ![icon](Customer Info) button on the toolbar to enter information about the customer. Only the customer's first name and last name are required. All other information is optional. When done, click on **OK**.

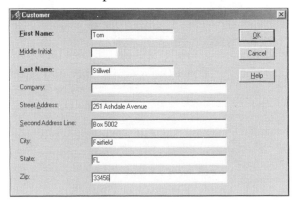

Fill in Customer Information.

2. **Job Name.** Click on the ![icon](Job Name) button on the toolbar. Then type the name of the job, such as *Room Addition*. When done, click on **OK**.

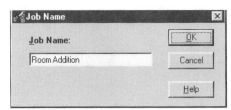

Enter a Job Name.

For transfers to QuickBooks — You can change the customer or job name after the file has been imported into QuickBooks. In QuickBooks, click on **Customers**. Click on **Customer Center**. Right-click on customer name or job name. Click on **Edit Customer:Job**. Then click on the tab of your choice.

3. Estimate Number. Click on the 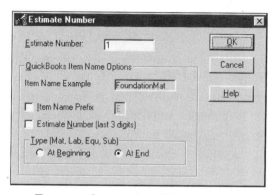 (Estimate Number) button on the toolbar. Job Cost Wizard keeps track of the last number used and recommends using the next number in sequence. Click on **OK** when done.

Check the Estimate Number.

For transfers to QuickBooks — When QuickBooks imports an estimate or invoice, subtotals in your estimate become cost category "items" in QuickBooks. By default, cost category names in QuickBooks are the first 28 characters of estimate subtotal names plus the work type, either *Mat*, *Lab*, *Equ*, or *Sub*. You can change this default in the Estimate Number dialog box.

Job Cost Wizard Prints Invoices Your Way

Your estimates should cover every cost in a job. But your bids and invoices don't have to show all the details and reveal your markup. So Job Cost Wizard gives you choices about showing or hiding the details and markup.

Amount of Detail

To set the amount of detail, click on the (Details) button.

If **Show All** is selected, every item in your bid or invoice will show a cost and each will become a cost category on the QuickBooks Item List. *Subtotals Only* is the default and will usually be a better choice.

If **Subtotals Only** is selected, subtotals will be the only costs in your bid or invoice. Each subtotal in your estimate becomes a cost category on the QuickBooks Items List. That's usually the best choice. If *Subtotals Only* is selected for all four cost categories, select **Omit Work Descriptions** to show subtotal categories but hide all work descriptions.

How much detail do you want to show?

The names you give to subtotals in National Estimator become cost category names in QuickBooks. Cost lines in National Estimator not followed by a subtotal become the "Project" subtotal.

If **Omit Entirely** is selected, neither costs nor descriptions will appear for that type of cost — either material, labor, equipment or total only (subcontract). Use Omit Entirely for materials, for example, when materials are being furnished by the owner.

"Total only" costs are assumed to be subcontract items. Subcontract items have a cost in the total column but no cost for material, labor, or equipment.

Markup and Tax

Use the three buttons at the bottom of the Amount of Detail dialog box to show or hide markup (overhead, contingency and profit) and tax.

In Every Cost Item distributes markup and tax proportionately throughout the estimate. There's no mention of overhead, profit or markup anywhere in the estimate or invoice.

At End of Estimate puts markup and tax at the end of the estimate, as in National Estimator.

Omit Entirely omits markup and tax from the estimate or invoice. Use this option if you prefer to add markup and tax in QuickBooks.

Select **Show Only Item Totals** if you don't want the invoice or estimate to show any breakdown of material, labor or equipment costs.

Click on **OK** when done with the Amount of Detail dialog box.

QuickBooks Account Names

Estimates and invoices imported into QuickBooks include expense and income account names. If the imported accounts do not exist already in your QuickBooks company, QuickBooks will create new accounts. You can control the names of these accounts by making changes in the QuickBooks Options dialog box.

▌ Click on the 🖹 (QuickBooks Options) button on the toolbar, -or-

▌ Click on **Options** on the menu bar. Then click on **QuickBooks Options**.

Enter the names you prefer for Income, Cost of Goods, Markup and Tax accounts. Change "Material Tax" to "FL Sales Tax," for example, if the job is taxable under Florida law. QuickBooks will keep track of tax due in each state where you do business.

Click **Use "Contractors Guide" Accounts** if you prefer the account names recommended in *Contractor's Guide to QuickBooks Pro*. To restore the default account names, click **Reset**.

If QuickBooks Pro version 2002 or later is installed on the computer, you should see a check mark beside *Use qbXML to integrate with QuickBooks*. XML exports to QuickBooks, as will be explained on the next page. If QuickBooks Pro is installed but there is no check mark, click on **Use qbXML to integrate with QuickBooks**. You'll be asked to identify the QuickBooks Company to receive imports from Job Cost Wizard. Select the company file you prefer and click **Open**. See *Exporting an Estimate to QuickBooks* on the next page for more on opening the XML link to QuickBooks.

Click on OK when done with QuickBooks Options.

Set up for automatic operation.

Automatic

Job Cost Wizard always requires customer information, a job name and a job number before exporting an estimate. In automatic mode, Job Cost Wizard opens the Customer Info, Job Name and Estimate Number dialog boxes automatically after opening any estimate.

Job Cost Wizard runs in automatic mode when there is a check mark beside Automatic on the Options menu. To change to automatic mode, click on **Options** on the menu bar. Then click on **Automatic**.

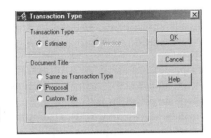

Changing the Document Title to "Proposal."

Transaction Type and Your Company Info

On the Options menu, click **Transaction Type** and select *Estimate* or *Invoice.* Document Title is either *Same as Transaction Type, Proposal* or your own *Custom Title.* For QuickBooks exports, the transaction type has to be either Estimate or Invoice.

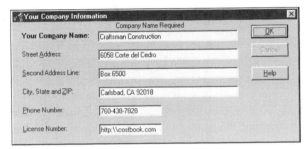

Change your company name or address.

On the Options menu, click **Your Company Info** to change the company name or address.

Click on the **Print** icon or **File** and **Print**. Then click **OK** to print the document.

Exporting an Estimate to QuickBooks

If you have QuickBooks 2002 or later, Job Cost Wizard will select XML export by default. This is the preferred way to export. If you have an earlier version of QuickBooks, exports will create an Intuit Interface File (IIF). XML exports happen over a direct link between the two programs. For IIF exports, Job Cost Wizard writes a file in the QuickBooks folder.

If you have 2002 QuickBooks or later, use the XML export:

1. Begin by clicking the QB icon.

2. Job Cost Wizard will advise that QuickBooks is asking permission to access the company file.

3. If you receive a warning about security level, follow instructions on the screen.

4. Click **Launch QuickBooks** and select the company file to receive the export.

5. QuickBooks will open.

6. Click **Yes, Always** to grant Job Cost Wizard access to the QuickBooks company file.

7. A bar will report that the transfer is in progress. Click **OK** when the export is successful.

8. In QuickBooks, click **Customers**. Click **Create Estimates**. Then click **Previous** to see the imported estimate. You can skip information on IIF exports and go right to *Turn an Estimate into an Invoice.*

If you have an older version of QuickBooks, *use the IIF export:*

QB icon.

1. Begin by clicking the QB icon.

2. Change the drive or the folder if the QuickBooks folder is not listed at the right of Save in.

3. Check the file name to be sure it is what you want.

4. If you use QuickBooks Pro, click on **Estimate**.

5. If you use regular QuickBooks, click on **Invoice**.

6. When complete, click on **Save**.

Exporting an estimate to QuickBooks Pro does not affect the original estimate in any way. Job Cost Wizard can open and export an estimate as many times as you want. To create a second copy of the same estimate with different Job Cost Wizard options, save with a slightly different file name. But note that you can import an estimate for any customer and job only once. Details are on the next page.

Exporting as an estimate to QuickBooks Pro.

Opening IIF Export Files in QuickBooks

Importing into QuickBooks.

Once an estimate or invoice has been written to file with Job Cost Wizard, start QuickBooks:

1. Click on **File**.

2. Click on **Utilities**.

3. Click on **Import**.

4. Click on **IFF Files**.

5. Browse to the location where the file was saved.

6. Double-click on the name of the estimate or invoice you want.

7. Click on **OK** when the import is complete.

8. Click on **Customers**.

9. Click on **Create Estimates** if you saved the file as an estimate.

10. Click on **Create Invoices** if you saved the file as an invoice.

11. Click on **Previous** to see the file just imported.

Important Note! With QuickBooks Pro 2000 and higher, you can import an estimate only once from an IIF. On second import of the same estimate, you'll see an error message, "Can't record invalid transaction." If you make a mistake and want to import an estimate again, delete the previous imported estimate before importing again. Instructions for deleting an estimate begin under *Don't Worry About Making a Mistake*. If you want two versions of any one estimate in QuickBooks Pro, save the alternate estimate with a slightly different job name or customer name. That makes the estimate different enough so it will import perfectly into QuickBooks.

Cost	Amount	Markup	Total
3,544.38	3,544.38		3,544.38
3,924.38			3,924.38
			7,468.76

Forcing a figure into the "Amount" column.

Filling in The "Amount" Column (IIF only)

An imported estimate is not complete until some figure appears in the Amount column of each cost line. So long as the Amount column is blank, QuickBooks will consider the estimated cost for that line to be zero (even when numbers appear in the Total column). If the Amount column for an entire estimate is left blank, QuickBooks reports will show the estimated cost for that job to be zero.

The fastest way to fill in the Amount column for a QuickBooks estimate is to click on a number in the "Cost" column and change the figure by a penny. Continue clicking and changing costs by a penny until every row of costs includes a figure in the Amount column.

Turn an Estimate into an Invoice in QuickBooks

First, decide if you want an invoice for the whole job or for just part of the job (progress billing). If you prefer progress billing:

1. Click on **Edit**, click on **Preferences**, click on **Jobs & Estimates**, click on **Company Preferences**, click on **Yes** under **Do You Do Progress Invoicing?** Click on **OK**.

To create the invoice:

1. Click on the **Create Invoice** button at the top of the QuickBooks estimate screen.

2. Click on **Yes** to record changes to the estimate.

3. If you selected progress billing, enter a percentage or select items to be invoiced.

4. Click on **OK** and QuickBooks creates the invoice.

5. Make changes to the invoice if you want.

6. Click to print one copy for your file and another for your customer.

7. Click on **Save & Close** when done. (Processing the file may take a little time.)

QuickBooks reports will now include totals from the job just invoiced.

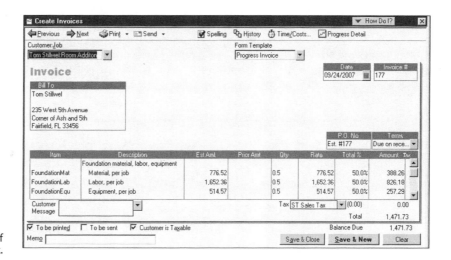

An invoice for the first half of the foundation work.

Don't Worry About Making a Mistake

It's easy to delete an estimate.

QuickBooks is very forgiving. Practice all you want. Experiment any way you want. Then delete any estimate or invoice to remove every trace of it from QuickBooks. With the offending estimate or invoice displayed:

1. Click on **Edit**.

2. Click on **Delete Estimate** (or Invoice).

3. Click on **OK**.

4. The estimate (or invoice) is deleted.

Your Jobs in QuickBooks

▌ Click on **Reports, Company & Financial, Profit & Loss By Job** to see job income and expense.

▮ Click on **Reports**, **Company & Financial**, **Balance Sheet Standard** to see the new receivables total.

▮ Click on **Reports**, **Jobs, Time & Mileage**, **Job Estimates vs. Actuals Detail** and select the **Customer:Job** to see a detailed cost comparison for the job. Until you start paying bills, the Actual Cost column will be all zeros.

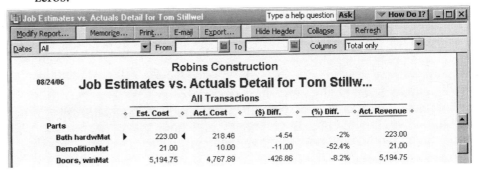

Job Estimates vs. Actuals Detail.

Paying Bills by Cost Category

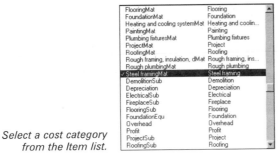

Select a cost category from the Item list.

When you pay vendors and subcontractors the amount paid is charged to the job and deducted from your bank balance.

1. Click on **Banking**.

2. Click on **Write Checks**.

3. Fill in the **Pay to the Order** line.

4. Click on the **Items** tab.

5. Click on the down triangle in the Item column of the check stub to open the Item list.

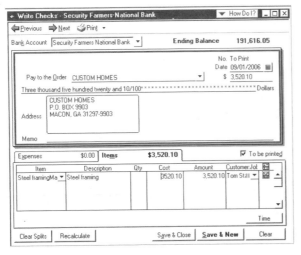

Paying a bill for the steel framing on the Stillwel job.

6. Select a *Mat*, *Sub* or *Equ* cost category from the Item list. These are the material, subcontract and equipment subtotal costs from your estimate.

7. Click in the amount column and enter the amount paid for the item described. Click on the down triangle under Customer:Job and select the correct customer and job.

8. This amount is not billable to your customer. So click on the icon representing an invoice to put a red X over the icon.

9. One check can cover items in several cost categories and even costs on several jobs. Click again on the next line down in the Item column and find the next cost item.

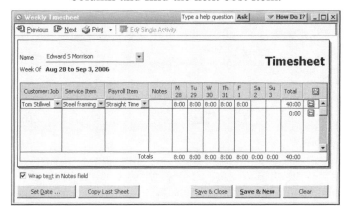

Fill out the timesheet by customer, service and payroll item.

Creating Payroll by Cost Category

When you write payroll checks, the amount paid is charged to the job and deducted from your bank balance.

1. Click on **Employees**.

2. Click on **Enter Time**.

3. Click on **Use Weekly Time Sheets**.

4. Click on the down triangle opposite Name and select the employee to be paid.

5. On the Timesheet, click on the down arrow under Customer:Job and select the first job where the employee worked.

Under Service Item, select a labor cost category from the list.

6. Under Service Item, select a Lab cost category from the list. These are labor subtotal costs from your estimate.

7. Under Payroll Item, select the type of pay, such as straight time.

8. Enter the number of hours worked for that cost category.

9. Any timesheet can cover work done on several jobs and many service items.

10. When finished recording time for an employee, click on **Save & Close**.

To actually produce paychecks:

1. Click on **Employees**.

2. Click on **Pay Employees**.

3. Check the names of the employees to be paid.

4. When done, click on **Create**.

Index

Practical References for Builders

Building Contractor's Exam Preparation Guide

Passing today's contractor's exams can be a major task. This book shows you how to study, how questions are likely to be worded, and the kinds of choices usually given for answers. Includes sample questions from actual state, county, and city examinations, plus a sample exam to practice on. This book isn't a substitute for the study material that your testing board recommends, but it will help prepare you for the types of questions — and their correct answers — that are likely to appear on the actual exam. Knowing how to answer these questions, as well as what to expect from the exam, can greatly increase your chances of passing.
320 pages, 8½ x 11, $35.00

National Home Improvement Estimator

Current labor and material prices for home improvement projects. Provides manhours for each job, recommended crew size, and the labor cost for removal and installation work. Material prices are current, with location adjustment factors and free monthly updates on the Web. Gives step-by-step instructions for the work, with helpful diagrams, and home improvement shortcuts and tips from experts. Includes a CD-ROM with an electronic version of the book, and *National Estimator*, a stand-alone *Windows*™ estimating program, plus an interactive multimedia tutorial that shows how to use the disk to compile home improvement cost estimates. **520 pages, 8½ x 11, $58.75. Revised annually**

Contractor's Guide to the Building Code

Explains in plain, simple English just what the 2006 *International Building Code* and *International Residential Code* require. Building codes are elaborate laws, designed for enforcement; they're not written to be helpful how-to instructions for builders. Here you'll find down-to-earth, easy-to-understand descriptions, helpful illustrations, and code tables that you can use to design and build residential and light commercial buildings that pass inspection the first time. Written by a former building inspector, it tells what works with the inspector to allow cost-saving methods, and warns what common building shortcuts are likely to get cited. Filled with the tables and illustrations from the *IBC* and *IRC* you're most likely to need, fully explained, with examples to guide you. Includes a CD-ROM with the entire book in PDF format, and with an easy search feature.
408 pages, 8½ x 11, $66.75

National Repair & Remodeling Estimator

The complete pricing guide for dwelling reconstruction costs. Reliable, specific data you can apply on every repair and remodeling job. Up-to-date material costs and labor figures based on thousands of jobs across the country. Provides recommended crew sizes; average production rates; exact material, equipment, and labor costs; a total unit cost and a total price including overhead and profit. Separate listings for high- and low-volume builders, so prices shown are specific for any size business. Estimating tips specific to repair and remodeling work to make your bids complete, realistic, and profitable. Includes a CD-ROM with an electronic version of the book with *National Estimator*, a stand-alone *Windows*™ estimating program, plus an interactive multimedia video that shows how to use the disk to compile construction cost estimates.
328 pages, 8½ x 11, $58.50. Revised annually

Estimating & Bidding for Builders & Remodelers

This 5th edition has all the information you need for estimating and bidding new construction and home improvement projects. It shows how to select jobs that will be profitable, do a labor and materials take-off from the plans, calculate overhead and figure your markup, and schedule the work. Includes a CD with an easy-to-use construction estimating program and a database of 50,000 current labor and material cost estimates for new construction and home improvement work, with area modifiers for every zip code. Price updates on the Web are free and automatic. **272 pages, 8½ x 11, $89.50**

Contractor's Plain-English Legal Guide

For today's contractors, legal problems are like snakes in the swamp — you might not see them, but you know they're there. This book tells you where the snakes are hiding and directs you to the safe path. With the directions in this easy-to-read handbook you're less likely to need a $200-an-hour lawyer. Includes simple directions for starting your business, writing contracts that cover just about any eventuality, collecting what's owed you, filing liens, protecting yourself from unethical subcontractors, and more. For about the price of 15 minutes in a lawyer's office, you'll have a guide that will make many of those visits unnecessary. Includes a CD-ROM with blank copies of all the forms and contracts in the book.
272 pages, 8½ x 11, $49.50

Residential Construction Performance Guidelines

Created and reviewed by more than 300 builders and remodelers, this guide gives cut-and-dried construction standards that should apply to new construction and remodeling. It defines corrective action necessary to bring all construction up to standards. Standards are listed for sitework, foundations, interior concrete slabs, basement and crawl spaces for block walls and poured walls, wood-floor framing, beams, columns and posts, plywood and joists, walls, wall insulation, windows, doors, exterior finishes and trim, roofs, roof sheathing, roof installation and leaks, plumbing, sanitary and sewer systems, electrical, interior climate control, HVAC systems, cabinets and countertops, floor finishes and more.
120 pages, 6½ x 8½, $39.95. Published by NAHB Remodelers Council

Getting Financing & Developing Land

Developing land is a major leap for most builders — yet that's where the big money is made. This book gives you the practical knowledge you need to make that leap. Learn how to prepare a market study, select a building site, obtain financing, guide your plans through approval, then control your building costs so you can ensure yourself a good profit. Includes a CD-ROM with forms, checklists, and a sample business plan you can customize and use to help you sell your idea to lenders and investors.
232 pages, 8½ x 11, $39.00

How to Succeed With Your Own Construction Business

Everything you need to start your own construction business: setting up the paperwork, finding the work, advertising, using contracts, dealing with lenders, estimating, scheduling, finding and keeping good employees, keeping the books, and coping with success. If you're considering starting your own construction business, all the knowledge, tips, and blank forms you need are here.
336 pages, 8½ x 11, $28.50

Construction Estimating

This unusually well-organized book shows the best and easiest way to estimate materials for room additions or residential structures. It gives estimating tables and procedures needed to make a fast, accurate, and complete material list of the structural members found in wood and steel-framed buildings. This book is divided into 72 units, each of them covering a separate element in the estimating procedure. Covers estimating foundations, floor framing, wall framing, ceiling framing, roof framing, roofing materials, exterior and interior finish materials, hardware, steel joist floor framing, steel stud framing, and steel ceiling joist and rafter framing.
496 pages, 8½ x 11, $49.50

Contractor's State License Guide - 'B' General Building Exam

Everything you need to pass the General Building License exam for the State of California. You'll find hundreds of multiple choice questions and answers to help you pass the state exam. Originally written for the State of California, this exam guide can be used to help you raise your score on any state's licensing exam. Includes two books: a study guide and a sample test. **350 pages, 8½ x 11, $99.00**

Steel-Frame House Construction

Framing with steel has obvious advantages over wood, yet building with steel requires new skills that can present challenges to the wood builder. This new book explains the secrets of steel framing techniques for building homes, whether pre-engineered or built stick by stick. It shows you the techniques, the tools, the materials, and how you can make it happen. Includes hundreds of photos and illustrations, plus a FREE download with steel framing details, a database of steel materials and manhours, with an estimating program. **320 pages, 8½ x 11, $39.75**

Craftsman's Construction Installation Encyclopedia

Step-by-step installation instructions for just about any residential construction, remodeling or repair task, arranged alphabetically, from Acoustic tile to Wood flooring. Includes hundreds of illustrations that show how to build, install, or remodel each part of the job, as well as manhour tables for each work item so you can estimate and bid with confidence. Also includes a FREE download with all the material in the book, handy look-up features, and the ability to capture and print out for your crew the instructions and diagrams for any job. **792 pages, 8½ x 11, $65.00**

Contractor's Survival Manual Revised

This completely revised book explains how to survive hard times and succeed during the up cycles. Shows what to do when the bills can't be paid, finding money and buying time, transferring debt, and all the alternatives to bankruptcy. Explains how to build profits, avoid problems in zoning and permits, taxes, time-keeping, and payroll. Unconventional advice on how to invest in inflation, get high appraisals, trade and postpone income, and stay hip-deep in profitable work. **336 pages, 8½ x 11, $38.00**

Construction Forms & Contracts

125 forms you can copy and use — or load into your computer (from the FREE disk enclosed). Then you can customize the forms to fit your company, fill them out, and print. Loads into *Word for Windows, Lotus 1-2-3, WordPerfect, Works,* or *Excel* programs. You'll find forms covering accounting, estimating, fieldwork, contracts, and general office. Each form comes with complete instructions on when to use it and how to fill it out. These forms were designed, tested and used by contractors, and will help keep your business organized, profitable and out of legal, accounting and collection troubles. Includes a CD-ROM for *Windows*™ and *Mac*™.
432 pages, 8½ x 11, $41.75

The Contractor's Legal Kit

Stop "eating" the costs of bad designs, hidden conditions, and job surprises. Set ground rules that assign those costs to the rightful party ahead of time. And it's all in plain English, not "legalese." For less than the cost of an hour with a lawyer you'll learn the exclusions to put in your agreements, why your insurance company may pay for your legal defense, how to avoid liability for injuries to your sub and his employees or damages they cause, how to collect on lawsuits you win, and much more. It also includes a FREE computer disk with contracts and forms you can customize for your own use. **352 pages, 8½ x 11, $69.95**

Builder's Guide to Accounting Revised

Step-by-step, easy-to-follow guidelines for setting up and maintaining records for your building business. This practical guide to all accounting methods shows how to meet state and federal accounting requirements, explains the new depreciation rules, and describes how the Tax Reform Act can affect the way you keep records. Full of charts, diagrams, simple directions and examples, to help you keep track of where your money is going. Recommended reading for many state contractor's exams. Each chapter ends with a set of test questions, and a CD-ROM included FREE has all the questions in interactive self-test software. Use the Study Mode to make studying for the exam much easier, and Exam Mode to practice your skills. **356 pages, 8½ x 11, $35.50**

Estimating With Microsoft *Excel*

Most builders estimate with *Excel* because it's easy to learn, quick to use, and can be customized to your style of estimating. Here you'll find step-by-step instructions on how to create your own customized automated spreadsheet estimating program for use with *Excel*. You'll learn how to use the magic of *Excel* to create detail sheets, cost breakdown summaries, and links. You'll put this all to use in estimating concrete, rebar, permit fees, and roofing. You can even create your own macros. Includes a CD-ROM that illustrates examples in the book and provides you with templates you can use to set up your own estimating system. **148 pages, 7 x 9, $39.95**

CD Estimator

If your computer has *Windows*™ and a CD-ROM drive, CD Estimator puts at your fingertips over 150,000 construction costs for new construction, remodeling, renovation & insurance repair, home improvement, framing & finish carpentry, electrical, concrete & masonry, painting, earthwork & heavy equipment and plumbing & HVAC. Monthly cost updates are available at no charge on the Internet. You'll also have the *National Estimator* program — a stand-alone estimating program for *Windows*™ that *Remodeling* magazine called a "computer wiz," and *Job Cost Wizard*, a program that lets you export your estimates to QuickBooks Pro for actual job costing. A 60-minute interactive video teaches you how to use this CD-ROM to estimate construction costs. And to top it off, to help you create professional-looking estimates, the disk includes over 40 construction estimating and bidding forms in a format that's perfect for nearly any *Windows*™ word processing or spreadsheet program.
CD Estimator is $78.50

Markup & Profit: A Contractor's Guide

In order to succeed in a construction business, you have to be able to price your jobs to cover all labor, material and overhead expenses, *and* make a decent profit. The problem is knowing what markup to use. You don't want to lose jobs because you charge too much, and you don't want to work for free because you've charged too little. If you know how to calculate markup, you can apply it to your job costs to find the right sales price for your work. This book gives you tried and tested formulas, with step-by-step instructions and easy-to-follow examples, so you can easily figure the markup that's right for your business. Includes a CD-ROM with forms and checklists for your use. **320 pages, 8½ x 11, $32.50**

National Contractor's Exam Study Guide

This book is not state-specific, but it is code-specific. It's guaranteed to help you obtain the knowledge and confidence needed to pass your contractor's exam on the first try. It lists 1,500 questions and answers — presented in the same format used on the actual exam — plus numerous references to the 2006 *International Building Code*. You'll find detailed illustrations that help clarify complicated code issues. And it also shows how to use your local codebooks to answer exam questions. Covers administration, definitions, use and occupancy classifications, special detailed requirements based on use and occupancy, general building heights and areas, types of construction, fire-resistant-rated-construction, interior finishes, fire-protection systems, means of egress, accessibility, energy efficiency, exterior walls, roof assemblies and rooftop structures, structural design and tests, soils and foundations, concrete, special construction, encroachments into public right-of-way, safeguards during construction, existing structures and more. **359 pages, 8½ x 11, $39.95**

Contractor's Guide to Change Orders

This book gives you the ammunition you need to keep contract disputes from robbing you of your profit. You'll learn how to identify trouble spots in your contract, plans, specifications and site; negotiate and resolve change order disputes, and collect facts for evidence to support your claims. You'll also find detailed checklists to organize your procedures, field-tested sample forms and worksheets ready for duplication, and various professional letters for almost any situation. **382 pages, 8½ x 11, $79.00**

Handbook of Construction Contracting

Volume 1: Everything you need to know to start and run your construction business; the pros and cons of each type of contracting, the records you'll need to keep, and how to read and understand house plans and specs so you find any problems before the actual work begins. All aspects of construction are covered in detail, including all-weather wood foundations, practical math for the job site, and elementary surveying. **416 pages, 8½ x 11, $32.75**

Volume 2: Everything you need to know to keep your construction business profitable; different methods of estimating, keeping and controlling costs, estimating excavation, concrete, masonry, rough carpentry, roof covering, insulation, doors and windows, exterior finishes, specialty finishes, scheduling work flow, managing workers, advertising and sales, spec building and land development, and selecting the best legal structure for your business. **320 pages, 8½ x 11, $33.75**

JLC Field Guide to Residential Construction

Volume 1: The ultimate visual quick-reference guide for construction professionals. Over 400 precisely-detailed drawings with clear concise notes and explanations that show you everything from estimating and selecting lumber to foundations, roofing, siding and exteriors. Explains code requirements for all U.S. building codes. Spiral bound. **386 pages, 8½ x 11, $69.95**

Volume 2: Over 300 detailed technical drawings, illustrations and tables giving key dimensions and critical details from accessible kitchens to zoned heating systems. This new version of JLC Field Guide has the field-proven principles, methods and materials on all facets of interior construction and remodeling. Spiral bound. **350 pages, 8½ x 11, $69.95**

Smart Business for Contractors

In this book, a construction attorney explains how you should charge for your work, how to figure your overhead expenses, and how to calculate a realistic hourly rate to apply on each estimate. Includes how to bill and collect on your invoices, what you should always include in your contracts, and creative new ways of dealing with contract disputes. Shows how to keep customers happy so they'll hand you referrals, how best to handle subcontractors, and how to find a good accountant. You'll learn the pros and cons of incorporating, how to handle tax issues such as what you can and can't deduct and what you're allowed to depreciate, and how to plan the future of your company. Reading this book is like getting good advice from a construction lawyer — at a fraction of the cost. Published by: Taunton Press. **204 pages, 8½ x 11, $19.95**

Managing the Small Construction Business

Overcome your share of business hassles by learning how 50 small contractors handled similar problems in their businesses. Here you'll learn how they handle bidding, unit pricing, contract clauses, change orders, job-site safety, quality control, overhead and markup, managing subs, scheduling systems, cost-plus contracts, pricing small jobs, insurance repair, finding solutions to conflicts, and much more. **328 pages, 8½ x 11, $34.95**

JLC Guide to Moisture Control

Water can be a building's worst enemy, which is why we try so hard to keep it out of our homes and living spaces. The editors of the *Journal of Light Construction* have brought together leading experts to write this book on controlling practically all forms of moisture damage in residential buildings. It explains the mysteries of vapor barriers, provides innovative solutions for keeping basements and crawl spaces dry, and shows foolproof methods for flashing windows and doors so that they won't let in storm water. Also gives storm-proof stucco details as well as tips for keeping paints and stain on exterior wood. Provides leak-free flashing techniques for both steep and low-slope roofs and plans for a low-cost whole-house ventilation that works incredibly well. **410 pages, 8½ x 11, $39.95**

Warranties for Builders & Remodelers

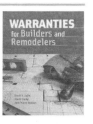

As customers become more knowledgeable, they expect warranties on your work. Providing a warranty also shows that you stand behind the work you do. But the question is: Are you doing it right, or are you digging yourself a hole you can fall into? This revised edition, written by the NAHB staff attorneys, helps you market your business by providing the services your customers require while protecting yourself and your business from liability. Here you'll find clear explanations of implied warranties, statutory warranties, warranties imposed by courts, and more. You'll find simple language and formats for warranty documents as well as tips for drafting warranties that really work. Each state has different legal requirements, so you'll also find a list of state-by-state cases and statutes applicable to construction claims. Warranties can be a two-edged sword. This book explains, in easy-to-understand language, your rights, responsibilities, and recommended practices and procedures to help make sure that you don't get cut. Published by: Builders Office Manuals. **130 pages, 6½ x 8½, $29.95**

Daily Log

Trying to keep track of all the small things that keep a construction company going can easily drive you to distraction — especially if you keep everything in your head. With this log book you can easily keep track of what happened each day on each job; the hours worked, the equipment rented, the material purchased, the craftsman on the job, the weather, and any immediate concerns worth noting. It has plenty of room to record important events, such as accidents. Good records can save your profit margin, as well as protect you from lawsuits. **416 pages, 8½ x 11, $59.95**

Drafting House Plans

Here you'll find step-by-step instructions for drawing a complete set of home plans for a one-story house, an addition to an existing house, or a remodeling project. This book shows how to visualize spatial relationships, use architectural scales and symbols, sketch preliminary drawings, develop detailed floor plans and exterior elevations, and prepare a final plot plan. It even includes code-approved joist and rafter spans and how to make sure that drawings meet code requirements. **192 pages, 8½ x 11, $34.95**

2003 *International Residential Code* Study Companion

Understand the 2003 *IRC*. To pass inspection, you have to build to Code. But hunting through the Code to find the rules, and then trying to understand them, can be a challenge. The 2003 *International Residential Code* Study Companion is guaranteed to make the job easier. It tells you the code section, quotes it, then explains it in plain English. Best of all, there's a large, clear illustration showing exactly how the work must be done. **374 pages, 8½ x 11, $54.00**

2008 *National Electrical Code*

This new electrical code incorporates sweeping improvements to make the code more functional and user-friendly. Here you'll find the essential foundation for electrical code requirements for the 21st century. With hundreds of significant and widespread changes, this 2008 *NEC* contains all the latest electrical technologies, recently developed techniques, and enhanced safety standards for electrical work. This is the standard all electricians are required to know, even if it hasn't yet been adopted by their local or state jurisdictions. **784 pages, 8½ x 11, $75.00**

Contractor's Law & Business for the California Contractor's Exam

This gigantic book is filled with everything you need to pass the California Contractor Law & Business exam: Lien law, workers' comp, CAL-OSHA requirements, California unemployment insurance act, state sales and use tax, business management, and basic bookkeeping and accounting practices. There are even 1,000 questions and answers included — just like the actual state tests, to make sure you pass with high marks. **402 pages, 8½ x 11, $99.95**

QuickBooks Help!

BuildWorks® is a specialized set of software tools that Synapse Software, Inc. has created and assembled into a business information management system for construction professionals.

The "BuildWorks System for Success" provides a manageable structure for information and work flow that can simplify the day-to-day operational tasks and profoundly improve your efficiency, productivity, and profitability.

BuildWorks offers you far more than the traditional construction software packages available today. BuildWorks is a complete construction information management system that provides you with a powerful "task based system" and a proven "work flow process" to track and manage your business operations with greater control, higher efficiency, and increased profitability. BuildWorks integrates the power of Microsoft Office™ with QuickBooks™ to provide a powerful, and easy-to-use business system created and priced specifically for today's competitive construction industry.

This innovative system is the perfect software solution for any builder or contractor who relies on subcontractors and who needs to accurately manage the day-to-day work of a productive and efficient staff. The short list of features in our HomeBuilder edition include job costing, construction estimating, project management, document management, scheduling, construction accounting, change orders, purchase & work orders, variance tracking, options & upgrades, construction draws, buyer selections, vendor bids, and specifications.

Choose from six value packed editions starting at only $295.

Mail This Card Today
For a Free Full Color Catalog

Over 100 books, annual cost guides and estimating software packages at your fingertips with information that can save you time and money. Here you'll find information on carpentry, contracting, estimating, remodeling, electrical work, and plumbing.

All items come with an unconditional 10-day money-back guarantee. If they don't save you money, mail them back for a full refund.

Name _____

Company _____

Address _____

City/State/Zip _____

Construction References **2008**
Handbooks for Professional Builders

2008 Cost Guides 2
Software 9
Estimating 10
Contractor Licensing 12
Electrical 16
Home Inspection 18
Remodeling 19
Building Code 20
Concrete 22
Construction 24
Carpentry 32
Office Manuals 35
Construction Law 40
Public Works 41
Mechanical 42
Painting 44
Land Development 44
Surveying & Heavy 45
Videos 47

Craftsman Book Company
6058 Corte del Cedro Carlsbad, CA 92011
www.craftsman-book.com

Craftsman Book Company / 6058 Corte del Cedro / P.O. Box 6500 / Carlsbad, CA 92018

ONLINE ACCOUNTING

P.O. Box 873

Eagle, ID 83616

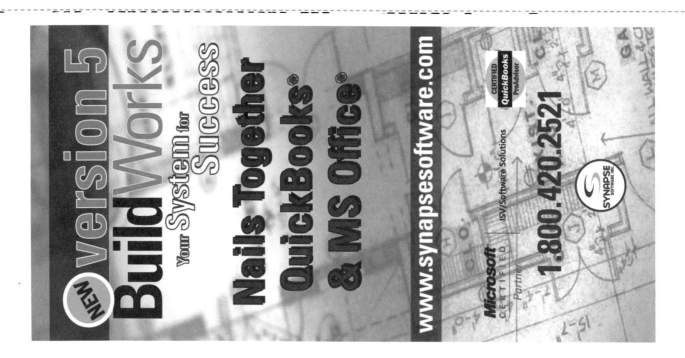

Download all of Craftsman's most popular costbooks for one
low price with the Craftsman Site License.
http://www.craftsmansitelicense.com

BUSINESS REPLY MAIL

FIRST CLASS MAIL PERMIT NO. 271 CARLSBAD, CA

POSTAGE WILL BE PAID BY ADDRESSEE

Craftsman Book Company
6058 Corte del Cedro
P.O. Box 6500
Carlsbad, CA 92018-9974